LAB WORKBOOK

Residential Wiring Concepts and Applications

Publisher
The Goodheart-Willcox Company, Inc.
Tinley Park, IL
www.g-w.com

ISBN 979-8-88817-007-6

1 2 3 4 5 6 7 8 9 – 24 – 28 27 26 25 24 23

Image Credits. Front cover: Klein Tools (left), JJ Gouin/Shutterstock.com (right), Lyudmila Zavyalova/Shutterstock.com (background image)

Contents

Name ______________________ Date ____________ Class ____________

CHAPTER 1 Getting Started in the Industry

Complete the following questions after carefully reading the corresponding chapter.

1. One reason that the electrical sector is stable is that electrical jobs cannot be _____ to other countries.

____

2. _____ *True or False?* It is best practice to have a résumé prepared *before* applying for a job.

3. List five work ethic skills critical to career success.

____

____

____

____

____

4. _____ *True or False?* The *NEC* was written to serve as an instructional manual for untrained persons.

5. _____ Which of the following is *not* one of the three categories of the *NEC*?
   A. Electrical licensing
   B. Explanatory material
   C. Mandatory rules
   D. Permissive rules

6. When states have an agreement where they will grant an electrical license to a resident of an adjoining state without testing it is called a(n) _____ agreement.

____

7. _____ *True or False?* If a homeowner adds a new circuit on their own home, an inspection is required.

8. _____ *True or False?* Homes across the US are powered by the 240/120-volt, three-phase, 3-wire system.

9. _____ In an electrical circuit, useful work is performed in the _____.
   A. grounded conductor
   B. load
   C. power source
   D. ungrounded conductor

10. _____ *True or False?* An ungrounded conductor carries electricity from the power source to the load.

11. Opposition to current flow is called _____.

____

12. _____ A 10-volt power source is placed across a 10 Ω load. What is the current?
    A. 1 A
    B. 10 A
    C. 100 A
    D. 1000 A

13. _____ How much power is used by a 30-volt load that produces a current of 6 amps?

A. 0.2 W

B. 5 W

C. 36 W

D. 180 W

14. An electric appliance is rated at 1800 watts and is connected to 240 volts. How much current will the machine draw?

____

Name ______________________________ Date ______________ Class __________

CHAPTER 2 Safety in Residential Wiring

Complete the following questions after carefully reading the corresponding chapter.

1. _____ *True or False?* The best practice is to clean the jobsite once a week.
2. _____ *True or False?* OSHA has a mission to ensure safe and healthful conditions in the workplace.
3. _____ Over one-third of all workplace fatalities come from _____.
   A. electrocution
   B. falls
   C. falling objects
   D. roll overs
4. _____ *True or False?* The level of current that can cause muscular contractions or muscular lock is from 10–20 mA.
5. _____ *True or False?* GFCI protection can be put into outlets, breakers, and extension cords.
6. Tamper-resistant receptacles contain a(n) _____ mechanism that will reject an object from being inserted into a single slot.

   ____

7. What does LOTO stand for?

   ____

8. _____ The highest usable step on a stepladder is which step?
   A. Top cap
   B. Second-highest step from the top cap
   C. Third-highest step from the top cap
   D. Fourth-highest step from the top cap
9. _____ A fire involving flammable liquids is a class _____ fire.
   A. A
   B. B
   C. C
   D. D
10. _____ A dry powder fire extinguisher can only be used on which class of fire?
    A. A
    B. B
    C. C
    D. D
11. Eyewear that has been tested and meets high-velocity impact requirements is marked as _____.

    ____

12. EH-rated shoes have thick soles that provide protection for up to _____ volts in dry conditions.

    ____

13. List four symptoms of heat exhaustion.

Name ______________________________ Date ______________ Class ____________

CHAPTER 3 Hand Tools

Complete the following questions after carefully reading the corresponding chapter.

1. Most accidents caused by hand tools are a result of misuse or improper _____.

____

2. _____ The two most common types of screwdrivers are _____.
   A. flathead and Phillips
   B. hex and flathead
   C. Phillips and Torx®
   D. Torx® and hex

3. Flathead screwdrivers are sized by the tip's actual _____.

____

4. _____ Which of the following types of pliers is best for cutting cables in tight spaces?
   A. Diagonal cutting pliers
   B. End-cutting pliers
   C. Long-nose pliers
   D. Pump pliers

5. _____ *True or False?* End-cutting pliers have a gap of approximately 1/16″ to prevent excessive damage to conductors in the jaws' grip.

6. _____ What type of wrench has a jaw that can be moved by a thumbscrew?
   A. Adjustable
   B. Box
   C. Hex
   D. Ratchet set

7. _____ *True or False?* Damage can occur from using a metric wrench on an SAE nut.

8. _____ *True or False?* A jab saw is also known as a keyhole saw.

9. _____ A saw blade with more teeth per inch is more _____ than a blade with fewer teeth.
   A. aggressive
   B. antagonistic
   C. intrusive
   D. invasive

10. _____ _____ are objects with threaded ends that allow the individual pieces to be coupled into a single, longer length.
    A. Cat's paws
    B. Fish sticks
    C. Fish tapes
    D. Tape measures

11. _____ When an item is plumb, it is _____ with the floor or ground.
    A. above
    B. below
    C. parallel
    D. perpendicular

12. _____ *True or False?* Prybars are smaller and shorter than crowbars.

13. _____ *True or False?* The condition you leave your jobsite in does *not* have a lasting impression on the customer.

Name ______________________ Date ____________ Class ____________

CHAPTER 4 Power Tools

Complete the following questions after carefully reading the corresponding chapter.

1. All 120-volt corded power tools should be plugged into _____ receptacles at a residential construction site.

   ____

2. _____ *True or False?* An advantage of cordless tools is that they offer less downtime than corded tools.

3. _____ *True or False?* It is safe to remove the blade guard on a circular saw when necessary.

4. _____ The most common power tool used in electrical work is the _____.
   A. drill
   B. hole saw
   C. jigsaw
   D. oscillating tool

5. Most general-use drills have a(n) _____ grip.

   ____

6. _____ *True or False?* The impact driver is used to drive or remove screws.

7. _____ The part of a drill bit that has flutes is called the _____.
   A. body
   B. clutch
   C. point
   D. shank

8. The most common rod-type drill bit is the _____ bit.

   ____

9. _____ Which type of bit has a flared, tungsten-carbide cutting tip?
   A. Auger
   B. Brad-tip
   C. Masonry
   D. Paddle

10. _____ *True or False?* Hole saws can only be used to cut through wood, never metal.

11. _____ *True or False?* It is recommended to drill metal at high speeds.

Notes

Name ______________________________ Date ______________ Class __________

CHAPTER 5 Electrical Test Equipment

Complete the following questions after carefully reading the corresponding chapter.

1. _____ *True or False?* A solenoid voltage tester does not need to be connected to exposed energized components to get a reading.

2. All measurements on a multimeter are made using test _____.

   ____

3. _____ Which of the following measurements should *not* be taken on an energized circuit?
   A. Continuity
   B. Current
   C. Voltage
   D. None of the above.

4. _____ *True or False?* The clamp meter is a safer alternative to the in-line ammeter.

5. _____ If a conductor is wrapped around a clamp meter twice and the clamp meter reads 24 amps, what is the actual current running through the conductor?
   A. 6 amps
   B. 12 amps
   C. 24 amps
   D. 48 amps

6. Some brands of clamp meter use _____ instead of spring-loaded clamps.

   ____

7. _____ *True or False?* You should *never* plug a receptacle tester into an energized receptacle.

8. _____ Which of the following is continuity testing *not* used for?
   A. Determining the condition of a fuse
   B. Finding a circuit breaker
   C. Testing the functionality of switches
   D. Tracing conductors

9. Continuity tests the circuit for completeness using the _____ function of a multimeter.

   ____

10. An open switch has _____ resistance.

    ____

Notes

Name ______________________ Date ______________ Class __________

CHAPTER 6 Fasteners and Anchors

Complete the following questions after carefully reading the corresponding chapter.

1. _____ *True or False?* Coarse-thread screws do not use a matching nut. They make their threading as they are driven into a material.

2. _____ Which of the following screws has the largest diameter?
   A. 6-32 × 1
   B. 6-32 × 5
   C. 8-24 × 4
   D. 10-24 × 4

3. _____ In a screw with the callout 6-32 × 2, the number of threads per inch is _____.
   A. 2
   B. 6
   C. 32
   D. 62

4. _____ *True or False?* The higher the gauge number, the larger the diameter of the screw.

5. The screw size used to attach switches to receptacles is _____.

   ____

6. _____ An 8-32 screw is commonly used _____.
   A. to install switches and receptacles and their cover plates
   B. to secure luminaires to the outlet box
   C. as the green grounding screw in metallic boxes for bonding purposes
   D. to secure ceiling fans up to 70 pounds to the outlet box

7. _____ *True or False?* Machine screws have more space between each thread than coarse-thread screws.

8. A drywall screw is another name for a(n) _____ screw.

   ____

9. _____ A self-drilling screw will penetrate sheet metal easily without the need for a(n) _____ hole.

   ____

10. _____ For use in thicker metals, the _____ screw will make its own hole.
    A. machine
    B. self-drilling
    C. self-tapping
    D. sheetrock

11. _____ *True or False?* Items installed with molly bolts can be installed and taken down repeatedly with no loss of strength.

12. _____ Which type of screw has double threads to cut into the material and provide high holding power?
    A. concrete
    B. machine
    C. self-tapping
    D. sheetrock

13. Nails with a rounded, dimpled head designed to be counter-sunk below the surface are called _____.

    ____

Notes

Name ______________________ Date ____________ Class __________

CHAPTER 7 Electrical Boxes

Complete the following questions after carefully reading the corresponding chapter.

1. _____ *True or False?* Metallic boxes are used far more than nonmetallic boxes to install switches and receptacles.
2. _____ Which of the following boxes is often referred to as a nail-on box?
   A. Handy
   B. Metallic
   C. Nonmetallic
   D. Old work
3. Most nonmetallic boxes are made of _____ plastic.

   ____
4. The process of removing the sides of metallic boxes to join them together is called _____.

   ____
5. _____ Which of the following boxes is often referred to as a utility box?
   A. Handy
   B. Metallic
   C. Nonmetallic
   D. Old work
6. _____ *True or False?* Traditional round nail-on boxes are sturdier than ceiling fan-rated boxes.
7. _____ Boxes meant for installations where the drywall has already been added are called _____ boxes.
   A. handy
   B. new work
   C. old work
   D. shallow pan
8. _____ *True or False?* A damp location is defined as any area that is subject to saturation or is directly impacted by the weather.
9. Weatherproof boxes are made of cast _____.

   ____
10. _____ *True or False?* Raised covers are used for surface-mounted boxes.
11. _____ _____ covers have a spring-loaded flap that protects the receptacle when not in use and are for use in damp locations.
    A. Paster
    B. Raised
    C. Weatherproof
    D. Weatherproof while in use

Notes

Name ______________________ Date ____________ Class ____________

CHAPTER 8 Electrical Conductors and Connectors

Complete the following questions after carefully reading the corresponding chapter.

1. _____ Which of the following wires is the smallest?
   A. 16 AWG
   B. 14 AWG
   C. 6 AWG
   D. 4/0 AWG

2. What is the area in circular mils of a 12 AWG wire?

   ____

3. _____ Which two Type NM cables have the same-colored outer sheathing?
   A. 14 AWG and 12 AWG
   B. 12 AWG and 10 AWG
   C. 10 AWG and 8 AWG
   D. 8 AWG and 6 AWG

4. _____ *True or False?* An ungrounded, hot conductor can have red insulation.

5. _____ *True or False?* Both Type SEU and Type SER cables are suitable for underground use.

6. What does the *R* in *SER* stand for?

   ____

7. _____ In cable systems, sizes 8 AWG and 6 AWG have _____ insulated conductors.
   A. bare
   B. grounded
   C. solid
   D. stranded

8. _____ is more flexible than solid wire.

   ____

9. The air temperature surrounding conductors is called the _____ temperature.

   ____

10. According to *NEC Table 310.16*, what is the ampacity of an aluminum 12 AWG conductor with Type SA insulation?

    ____

11. _____ According to *NEC Table 310.16*, what is the ampacity of a copper 4 AWG conductor with Type THW insulation?
    A. 65
    B. 75
    C. 85
    D. 95

12. What AWG size copper conductor with Type SA insulation would be sufficient to power an 80-amp load?

    ____

13. _____ *True or False?* Aluminum expands and contracts to a greater degree than copper.

14. _____ *True or False?* A wire nut can accommodate up to seven wires.

15. Overtightening a cable connector can damage a cable and cause a(n) _____ circuit.

Name ______________________ Date ____________ Class ____________

CHAPTER 9 Electrical Conduit

Complete the following questions after carefully reading the corresponding chapter.

1. The primary purpose of electrical conduit is the physical _____ of conductors.

   ____

2. _____ *True or False?* The inner diameter determines the number of conductors a conduit can safely accommodate.
3. _____ Which of the following conduit types provides the most protection for conductors?
   A. Electrical metallic tubing
   B. Intermediate metal conduit
   C. Liquidtight flexible nonmetallic conduit
   D. Rigid polyvinyl chloride conduit
4. RMC and IMC are bent using _____ benders.

   ____

5. _____ *True or False?* All sizes of electrical metallic tubing can be easily cut with a hacksaw.
6. _____ *True or False?* Rigid polyvinyl chloride conduit is often used in underground installations.
7. _____ Which type of conduit comes with a coupling on one end called the bell end?
   A. Electrical metallic tubing
   B. Intermediate metal conduit
   C. Rigid polyvinyl chloride conduit
   D. Rigid metal conduit
8. _____ PVC must be securely fastened within _____ feet of any conduit termination.
   A. 3
   B. 5
   C. 6
   D. 10
9. _____ *True or False?* A bushing is a conduit fitting used to join two pieces of conduit together.
10. _____ The fitting used to join a conduit to a junction box is called a _____.
    A. bushing
    B. connector
    C. coupling
    D. terminal adapter
11. _____ *True or False?* Conduit bodies are available for both metallic and PVC conduits.

Notes

Name ______________________________ Date ______________ Class __________

CHAPTER 10 Electrical Devices

Complete the following questions after carefully reading the corresponding chapter.

1. _____ *True or False?* Luminaires and ventilation fans are outlets.

2. The most common devices installed in a residence are duplex _____.

 ____

3. _____ The grounded conductor is connected to the _____ on the face of a receptacle.
 A. grounded terminal screw
 B. horseshoe-shaped slot
 C. long slot
 D. short slot

4. _____ A 120-volt duplex receptacle with a T-shaped grounded slot has a _____-amp rating.
 A. 10
 B. 15
 C. 20
 D. 25

5. _____ *True or False?* A CU rated receptacle is rated for both copper and aluminum conductors.

6. GFCI receptacles prevent electrical shocks by _____ a circuit when a ground fault condition is detected.

 ____

7. _____ *True or False?* GFCI receptacles protect against shock when a user contacts both the grounded and the ungrounded circuit conductors simultaneously.

8. _____ *True or False?* Surface-mounted range receptacles require a box for installation.

9. *Section 210.70* requires at least _____ wall switch-controlled lighting outlet(s) to be installed in every habitable room.

 ____

10. _____ A _____ switch is always installed in pairs and controls loads from two wall locations.
 A. single-pole
 B. two-way
 C. three-way
 D. four-way

11. A switch that also allows the level of light emitted to be reduced is called a(n) _____ switch.

 ____

12. _____ *True or False?* Three-way dimmers are not available.

13. _____ When two devices are mounted on the same yoke, it is known as a _____ device.
 A. combination
 B. duplex
 C. pigtail
 D. three-way

Notes

Name ______________________ Date ____________ Class ____________

CHAPTER 11 Panelboards, Overcurrent Protective Devices, and Disconnects

Complete the following questions after carefully reading the corresponding chapter.

1. _____ *True or False?* A main breaker distribution panel can house fuses or circuit breakers.

2. With an MLO distribution panel, MLO stands for main lugs _____.

   ____

3. _____ *True or False?* A main breaker panelboard and main lugs only panelboard have the same installation requirements.

4. _____ The ampacity of the conductor must be _____ to the overcurrent device protecting it.
   A. equal
   B. greater or equal
   C. lesser or equal
   D. The two are not related.

5. A circuit breaker monitors the thermal properties of a circuit using a(n) _____ strip, a device made by fusing two dissimilar metals together.

   ____

6. _____ *True or False?* With an inverse-time breaker, the time delay increases as the magnitude of the current increases.

7. The level of fault current a circuit breaker can safely interrupt is called the _____ rating.

   ____

8. _____ Most residential circuit breakers have an AIC rating of _____ A.
   A. 1
   B. 100
   C. 10,000
   D. 1,000,000

9. _____ Tandem breakers are another name for _____ circuit breakers.
   A. double-pole
   B. GFCI
   C. single-pole
   D. twin

10. _____ *True or False?* Double-pole circuit breakers can be used for both 240-volt and 120/240-volt circuits.

11. A dual-element time-delay fuse uses a fusible link to protect against a high current and a(n) _____ link to protect against a sustained thermal overload condition.

    ____

12. _____ *True or False?* With a Type S fuse, any size fuse can be screwed into the fuse holder.

13. _____ In residential wiring, the most common type of disconnect for large electrical equipment is a _____.
    A. circuit breaker
    B. NEMA 3R
    C. plug
    D. pull-out

14. Disconnects that are installed outside and exposed to the weather must have a _____ rating.

    ____

Notes

Name ______________________________ Date ______________ Class __________

CHAPTER 12 Lamps and Lighting

Complete the following questions after carefully reading the corresponding chapter.

1. List three ways lights can be secured into a fixture.

____

____

____

2. _____ *True or False?* Cool light is more blue in color, and warm light is more yellow.

3. _____ A lamp's brightness level is measure in _____.
   A. Kelvins
   B. lumens
   C. luminous efficacy
   D. wattage

4. _____ *True or False?* Incandescent lamps typically have higher luminous efficacy than LEDs.

5. _____ A G40 lamp has a diameter of _____ inches.
   A. 4
   B. 5
   C. 8
   D. 40

6. On an incandescent bulb, what does the F shape designation stand for?

____

7. _____ *True or False?* A clear bulb emits a bright light and casts crisp, clean shadows.

8. In multi-bulb chandeliers, the base that is primarily used is the E12 ____ base.

____

9. _____ *True or False?* It is safe to operate an incandescent bulb at a voltage less than its rated voltage.

10. _____ A lamp that produces visible light through a reaction with the lamp's phosphor coating is a(n) _____ light.
    A. ballest
    B. fluorescent
    C. incandescent
    D. LED

11. _____ *True or False?* A ballast is only used to supply the high voltage required to start a fluorescent lamp.

12. _____ An LED lamp uses a(n) _____ to convert the incoming AC power into the DC power needed for operation.
    A. adapter
    B. ballest
    C. driver
    D. filament

13. Lighting used to illuminate a point of interest, such an architectural feature, is called _____ lighting.

____

Notes

Name ______ Date ______ Class ______

CHAPTER 13 Basic Switched Circuits

Complete the following questions after carefully reading the corresponding chapter.

1. In a complete circuit, an electrical current leaves the source, travels through the circuit, energizes a load, then returns to the _____.

______

2. _____ When using nonmetallic cable as the wiring method, current is supplied to the load with _____.
   A. the black ungrounded conductor
   B. the red ungrounded conductor
   C. the black or red ungrounded conductor
   D. neither the black nor red ungrounded conductor

3. _____ *True or False? Section 404.2(B)* permits switches to disconnect the grounded conductor.

4. _____ If a grounded conductor is required at a switch location, a _____-conductor cable is used as the switch loop.
   A. single
   B. two
   C. three
   D. four

5. The single-pole switch operates by opening and closing the circuit in _____ location.

______

6. _____ A single-pole switch has _____ terminal screws.
   A. one
   B. two
   C. three
   D. four

7. _____ *True or False?* In a three-way switched circuit, power can enter the circuit at either of the two switches.

8. The wiring scheme for a three-way switched circuit where the power and the three-conductor switch cables are installed at the lighting outlet location, and both three-wire conductors end in their respective boxes, is called the _____.

______

9. Four-way switches have four terminal screws, two for each pair of _____.

______

10. _____ *True or False?* The common conductor is *not* connected to the four-way switch.

Notes

Name ______________________ Date ____________ Class __________

CHAPTER 14 Advanced Switch Circuits

Complete the following questions after carefully reading the corresponding chapter.

1. _____ *True or False?* The *NEC* allows switched receptacles to serve in place of switched lighting outlets in bathrooms.

2. To split a receptacle, remove the _____ between the two brass-colored terminal screws of an ordinary duplex receptacle.

   ____

3. _____ *True or False?* A switched circuit using a single-pole switch can control more than one split receptacle.

4. _____ How many split receptacles can a switched circuit using three-way switches control?
   A. One
   B. Two
   C. Four
   D. As many as needed.

5. _____ When using a three-conductor cable to control split receptacles, common practice uses the _____ conductor for switched power.
   A. black
   B. red
   C. red or black
   D. white

6. When using a three-conductor cable to control split receptacles, the _____, grounded conductor provides the return path for the power.

   ____

7. _____ *True or False?* A constant-power receptacle always remains energized.

8. _____ *True or False?* When wiring a switch-controlled lighting outlet with a constant-power receptacle downstream, the black wire is connected to the switch and provides switched power to the lighting outlet.

9. A single power source can be _____ to provide power to different switched loads.

   ____

10. _____ *True or False?* When wiring a double-gang box with a constant power receptacle and single-pole switch-controlled light on a single feed, a two-conductor cable can be run from the switch location to the lighting outlet box.

Notes

Name ______________________ Date ____________ Class ____________

CHAPTER 15 The *NEC*: Purpose, Scope, and Navigation

Complete the following questions after carefully reading the corresponding chapter.

1. _____ *True or False?* The first edition of the *NEC* was published by the National Board of Fire Underwriters.
2. The *NEC* is published by the NFPA, which stands for _____.

   ____
3. The *NEC* is reviewed periodically by a team of authors divided into _____ Code-Making Panels.

   ____
4. _____ *True or False?* Anyone can submit proposals for changes to the *NEC.*
5. _____ *True or False? Article 90* states that the purpose of the *NEC* is to safeguard persons and property from the hazards arising from the use of electricity.
6. _____ Which of the following is covered by *NEC* standards?
   A. Automobiles
   B. Machine shops
   C. Railway cars
   D. Service laterals
7. The responsibility of enforcing the requirements of the *NEC* falls on the _____ (AHJ).

   ____
8. Actions that are allowed but not required in the *NEC* are _____ rules.

   ____
9. _____ In the *NEC*, which of the following are formatted in italics?
   A. Exceptions
   B. Informational notes
   C. Mandatory rules
   D. Permissive rules
10. _____ Which chapters in the *NEC* are the most useful for a residential electrician?
    A. Chapters 1 through 4
    B. Chapters 5 through 8
    C. Chapters 9 through 12
    D. All chapters are equally useful.
11. _____ In which chapter of the *NEC* can you find information about wiring for mobile and manufactured homes?
    A. Chapter 4
    B. Chapter 5
    C. Chapter 6
    D. Chapter 7
12. _____ *True or False?* The *NEC* outline format, from the most general to the most specific, is article, section, part, subdivision.
13. _____ *True or False?* Brackets used between the paragraphs indicate that one or more paragraphs have been deleted.
14. _____ *True or False?* When searching for information in the *NEC*, reviewing the Table of Contents for the most appropriate article is the best place to start.

Notes

Name ______________________ Date ____________ Class ____________

CHAPTER 16 Print Reading and Estimating

Complete the following questions after carefully reading the corresponding chapter.

1. On a blueprint, the relationship of the drawing size to the actual size is called _____.

____

2. _____ *True or False?* An architects' scale is an alternate name for an engineers' scale.

3. _____ *True or False?* Orthographic projection is a method of rendering a two-dimensional object in three dimensions.

4. _____ An orthographic projection of a single-story building generally results in _____ views of the building.
   A. two
   B. three
   C. four
   D. five

5. _____ The most basic floor plans, which tie together all other prints, are called the architectural _____.
   A. details
   B. drawings
   C. plans
   D. sections

6. Information such as the project name or owner and the name of the architect can be found in the _____ block.

____

7. Enlarged views of a portion of a building's plan that provide greater clarity are called architectural _____.

____

8. _____ Which of the following are identified by the reference letter S?
   A. Architectural drawings
   B. Architectural sections
   C. Site plans
   D. Structural plans

9. Architectural sections can be taken from floor plans and identified by section _____ lines.

____

10. A table that lists specific materials used in the building process is called a(n) _____.

____

11. _____ Written instructions that define the scope of work and the materials used in the construction process are called _____.
   A. details
   B. elevations
   C. sections
   D. specifications

12. _____ *True or false?* Solid lines are drawn from switch locations to the lighting outlets they control on an electrical plan.

13. _____ *True or false?* The first step in planning residential electrical design is to locate the breaker panel.

14. An estimate that involves counting and pricing every item to be supplied and installed is called a(n) _____ estimate.

____

Name ______________________________ Date ______________ Class __________

CHAPTER 17 Residential Framing

Complete the following questions after carefully reading the corresponding chapter.

1. _____ *True or False?* Lumber is strong in relation to its weight.

2. Wood that comes from slow-growing deciduous trees is called _____.

____

3. _____ A 2×4 piece of lumber has an actual measurement of _____.
   A. 1″ by 3″
   B. 1-1/2″ by 3-1/2″
   C. 2″ by 4″
   D. 2-1/2″ by 4-1/2″

4. Lumber processed with a water/copper solution that slows rot is called _____ treated lumber.

____

5. _____ The structural members that support the roof decking are the _____.
   A. dormers
   B. headers
   C. joists
   D. rafters

6. _____ In which type of framing is the first floor framed on top of the foundation and the subfloor installed on floor joists?
   A. Balloon framing
   B. Platform framing
   C. Post-and-beam framing
   D. Stick-built framing

7. In floor framing, the header is attached to the _____ plate, which is attached to the top of the foundation wall.

____

8. The portion of the wall frame that attaches to the subfloor is called the _____ plate, or bottom plate.

____

9. _____ *True or False?* Interior walls that run perpendicular to the floor and ceiling joists are often nonload-bearing walls.

10. _____ The vertical distance between the top plate of a wall and the ridge beam at the peak of a roof is called the _____.
    A. pitch
    B. rise
    C. run
    D. span

11. _____ Which of the following pitches corresponds to the *least* steep roof?
    A. 3/12
    B. 6/12
    C. 9/12
    D. 12/12

12. _____ The portion of the roof that overhangs the outer walls of the house is called a(n) _____.
A. dormer
B. eave
C. facia
D. valley

13. The front side of an eave is called the _____.

____

14. _____ *True or False?* Roof trusses are engineered roofing frames assembled off-site.

Name ______________________________ Date ______________ Class __________

CHAPTER 18 Branch Circuit/Receptacle Requirements

Complete the following questions after carefully reading the corresponding chapter.

1. The conductors between the final overcurrent protection device and the outlets being served are called a(n) _____ circuit.

____

2. _____ *True or False?* Large appliances like electric ranges might have their own branch circuit.

3. _____ A branch circuit that supplies only a single load is called a(n) _____ branch circuit.
   A. general-purpose
   B. individual
   C. single
   D. All branch circuits supply two or more loads.

4. A 20-amp bathroom branch circuit requires size _____ AWG conductors.

____

5. _____ What is the minimum number of 15-amp branch circuits required for a 2,000 square foot dwelling?
   A. Three
   B. Four
   C. Five
   D. Six

6. _____ *True or False?* The *NEC* allows a wall-switch-controlled lighting outlet or a switch-controlled receptacle for hallways.

7. _____ The *NEC* requires at least _____ lighting outlet(s) to be installed in attics where these spaces are used for storage.
   A. one
   B. two
   C. three
   D. No lighting outlets are required.

8. _____ The *NEC* allows floor receptacles installed within _____ inches of the wall to be counted as a required wall receptacle.
   A. 12
   B. 14
   C. 18
   D. 24

9. _____ The *NEC* requires each bathroom to have at least _____ receptacle outlet(s) installed within three feet of the outside edge of the sink.
   A. one
   B. two
   C. three
   D. There is no specific requirement.

10. _____ *True or False?* The *NEC* requires that single-family dwellings must have a minimum of four outdoor receptacles.

11. _____ The receptacle for a washing machine must be installed within _____ feet of the intended location of the machine.
A. 3
B. 6
C. 9
D. 12

12. _____ *True or False?* GFCI protection can be provided for a bathroom outlet by means of either a GFCI receptacle or a GCFI circuit breaker.

13. _____ Which of the following rooms is *not* required to have AFCI protection?
A. Bathroom
B. Bedroom
C. Closet
D. Dining room

Name ______________________ Date ______________ Class ____________

CHAPTER 19 Conductor Sizing and Ampacity Adjustments

Complete the following questions after carefully reading the corresponding chapter.

1. _____ *True or False?* The ampacity of a conductor can be affected by the number of current-carrying conductors run in the same raceway.

2. _____ Using *Table 310.16*, determine the allowable ampacity of a 10 AWG copper wire with type UF insulation.
   A. 20 amps
   B. 25 amps
   C. 30 amps
   D. 40 amps

3. _____ Using *Table 310.15(B)(1)(1)*, determine the ambient temperature correction factor of a conductor with a temperature rating of 75°C and an ambient temperature of 20°C.
   A. 1.05
   B. 1.08
   C. 1.11
   D. 1.15

4. Using *Table 310.15(B)(1)(1)* and *Table 310.16*, the corrected ampacity for a 10 AWG copper wire with type TW insulation and an ambient temperature of 35°C is _____ amps.

   ____

5. _____ Using *Table 310.15(C)(1)*, what is the adjustment factor for six 12-2 NM cables and three 12-3 NM cables that have been bundled together?
   A. 35 percent
   B. 40 percent
   C. 45 percent
   D. 50 percent

6. According to *Table 310.15(C)(1)*, the adjustment factor for two 12-2 NM cables and two 12-3 NM cables that have been bundled together is _____ percent.

   ____

7. _____ *True or False?* Longer runs of wire increase the voltage drop.

8. Voltage _____ on longer runs can be minimized by increasing the size of the conductors.

   ____

9. _____ *True or False?* The current draw of the load can affect the amount of voltage drop in a circuit.

10. The voltage drop in 100 feet of 4 AWG (41,740 cmil) 2 NM aluminum cable where the current is 100 amps is _____ volts.

    ____

11. The voltage drop over 300 feet of 2 NM cable is 27 volts. If the source voltage is 240 volts, the voltage drop as a percentage of the source is _____ percent.

    ____

12. _____ A circuit has 40 feet of 2 NM copper cable. The circuit is at 120 volts and 50 amps. Using *Table 8, Chapter 9*, what size conductor is necessary to keep the voltage drop under 5 percent? K for copper is 12.9.

A. 8
B. 10
C. 12
D. 14

Name ______________________________ Date ______________ Class __________

CHAPTER 20 Service Calculations

Complete the following questions after carefully reading the corresponding chapter.

1. If a load is expressed in watts or kilowatts, it is a nearly 100 percent _____ load.

____

2. _____ *True or False?* The major portion of the VA of a load is the inductive reactance component.

3. _____ The consumption of all electrical equipment drawing power collectively is the _____.
   A. calculated load
   B. connected load
   C. demand factor
   D. general load

4. _____ The method of service calculation that applies a single demand factor to the total connected load to determine the calculated load is called the _____ method.
   A. calculated
   B. connected
   C. general
   D. optional

5. _____ Using the general method, what is the estimated lighting load for a dwelling with 2,000 square feet of living space?
   A. 2000 VA
   B. 6000 VA
   C. 10,000 VA
   D. 12,000 VA

6. Using the general method, the estimated load coming from four small appliance circuits and one laundry circuit is _____ VA.

____

7. _____ *True or False?* When adding noncoincident loads to the service calculation, you should omit the smaller of the two loads.

8. _____ *True or False?* The minimum service size for a dwelling allowed by the *NEC* is 100 amps.

9. _____ *True or False?* With the optional method, as with the general method, the smaller of the heating or air-conditioning load is omitted from the calculation.

10. _____ While using *Table 220.55* to calculate loads for cooking appliances, which of the three columns simply applies a percentage reduction to the nameplate rating of the appliances?
    A. A and B
    B. A and C
    C. B and C
    D. A, B, and C

11. _____ *True or False?* The *NEC* does not permit reductions for multiple electric dryer loads.

12. _____ What is the demand factor on a dwelling with five dryers?
    A. 100 percent
    B. 85 percent
    C. 75 percent
    D. 25 percent

Notes

Name ______________________ Date ____________ Class ________

CHAPTER 21 Temporary Power

Complete the following questions after carefully reading the corresponding chapter.

1. _____ *True or False?* Temporary power service includes a meter enclosure.
2. _____ *True or False?* Temporary panels that provide power to a minimum of two ground-fault protected circuits to power receptacles are required by the *NEC*.
3. _____ *Section 590.4(D)(1)* prohibits the use of _____ on a circuit used for temporary lighting.
   A. AFCI protection
   B. GFCI protection
   C. receptacles
   D. Nothing is prohibited.
4. Service drop conductors drop from the utility transformer to the service _____.

   ____

5. All wood used for an overhead service mast should be _____ treated.

   ____

6. _____ The stakes that hold the diagonal support braces for an overhead service mast should be driven to a minimum depth of _____ feet to help to stabilize the mast.
   A. two
   B. four
   C. six
   D. eight
7. _____ The U in Type SEU cable stands for _____.
   A. underground
   B. ungrounded
   C. utility
   D. U-shaped
8. _____ *True or False?* Underground temporary services require the same extra bracing that an overhead service requires.
9. The service post for a temporary underground service should be at least _____ feet in the ground.

   ____

10. _____ *True or False?* Service laterals are conductors that supply underground service poles.

Notes

Name ______________________ Date ____________ Class ________

CHAPTER 22 Outlet Box Installation

Complete the following questions after carefully reading the corresponding chapter.

1. Which *NEC* article and part provides spacing guidance for the placement of receptacles in a dwelling?

____

2. _____ What is the lowest height at which a receptacle should be installed so that it does not interfere with a kitchen countertop and backsplash?
   A. 36 inches
   B. 38 inches
   C. 40 inches
   D. 42 inches

3. _____ *True or False?* The front edge of a nail-on box should be recessed no more than 1/2 inch from the finished surface if the wall finish is of noncombustible material.

4. _____ *True or False?* When installing a cut-in box, it is better to make the opening too small than too large.

5. _____ What is the maximum number of 14 AWG conductors that is permitted in an octagonal 4 × 1-1/4 metal box?
   A. 6
   B. 7
   C. 8
   D. 9

6. What is the maximum number of 18 AWG conductors that is permitted in a round 4 × 2-1/8 metal box?

____

7. _____ What is the minimum size metal device box required for five 16 AWG conductors?
   A. 3 × 2 × 1-1/2
   B. 3 × 2 × 2
   C. 3 × 2 × 2-1/4
   D. 3 × 2 × 2-1/2

8. _____ *True or False?* A single volume allowance of the largest conductor in the box is added for a hickey.

9. _____ The volume allowance for conductors less than 12 inches in length that pass through the box without a splice or connection is _____.
   A. 1/4 for each conductor
   B. one for each conductor
   C. two for each conductor
   D. None of the above.

10. _____ The volume allowance for pigtails is _____.
   A. 1/4 for each conductor
   B. one for each conductor
   C. two for each conductor
   D. None of the above.

11. The volume allowance required for three 18 AWG conductors and two 12 AWG conductors is _____ inches.

____

12. _____ What is the minimum size 4-inch square box that is required for three 16 AWG and four 14 AWG conductors?

A. 4 × 1-1/4
B. 4 × 1-1/2
C. 4 × 2-1/8
D. None of the above.

Name ______________________________ Date ______________ Class __________

CHAPTER 23 Conductor Routing and Installation

Complete the following questions after carefully reading the corresponding chapter.

1. _____ The primary type of wiring used in residential construction is _____.
   A. AC
   B. EMT
   C. MC
   D. NM

2. _____ *True or False?* The *NEC* requires the edge of a hole to be no less than 1-1/2 inches from the nearest edge of the framing member.

3. When sufficient space cannot be maintained from the edge of a stud, the *NEC* requires a _____ plate be installed to protect the conductors.

   ____

4. _____ Which section of the *NEC* requires all work to be executed in a neat manner?
   A. *Section 110.12*
   B. *Section 300.4(A)(1)*
   C. *Section 300.4(A)(2)*
   D. *Section 334.2*

5. _____ Branch circuits to the _____ can use 14 AWG conductors.
   A. bathroom
   B. foyer
   C. garage
   D. kitchen

6. _____ *True or False?* Type NMS is allowed to be embedded in concrete.

7. _____ *True or False?* An electrician should always estimate the cable length needed *before* securing the cable.

8. *Section 334.30* requires NM cable to be supported by staples, cable ties, straps, or hangers at intervals no greater than _____ inches.

   ____

9. _____ *True or False?* It is required by the *NEC* to strip the outer sheath of the wiring *after* the cable has been run into the box.

10. NM cable with a diameter of one inch should not have a bending radius less than _____ inch(es).

    ____

11. The *NEC* requires that all conductors extend at least _____ inches beyond the outer sheathing.

    ____

12. _____ *True or False?* The *NEC* requires that at least one inch of sheathing should extend into the outlet box beyond the clamping mechanism.

Notes

Name ____________________ Date ____________ Class ____________

CHAPTER 24 Service Equipment

Complete the following questions after carefully reading the corresponding chapter.

1. Once the overhead or underground service conductors are routed through the utility meter, they become service-_____ conductors.

____

2. Meter sockets that route the underground utility wiring from below grade to the meter terminals are often in a(n) _____ configuration.

____

3. _____ *True or False?* Meter sockets and service disconnecting means can be built from individual components or come as a preassembled combination.

4. The circuit conductors between the service equipment and the final branch circuit overcurrent protection device are called _____ conductors.

____

5. _____ How many conductors are found in a feeder?
   A. 2
   B. 3
   C. 4
   D. 5

6. A distribution panelboard that does *not* contain the service disconnecting means is known as a(n) _____-wire service.

____

7. A main bonding jumper can only be installed at the service _____.

____

8. _____ *True or False?* System grounding helps stabilize the voltage to the earth during normal operation.

9. _____ Any metal support structure in contact with the earth for no less than _____ feet vertically can be used as a grounding electrode.
   A. 5
   B. 10
   C. 20
   D. 40

____

10. _____ *True or False?* If a grounding electrode conductor has insulation, the *NEC* requires the insulation to be green or black.

11. _____ The *NEC* requires that a pipe electrode be installed where _____ feet of length is in direct contact with the earth.
   A. 4
   B. 6
   C. 8
   D. 10

12. _____ Which electrode is also known as the Ufer ground?
   A. Concrete-encased electrode
   B. Plate electrode
   C. Rod electrode
   D. Underground metal water pipe

Name ______________________________ Date ______________ Class __________

CHAPTER 25 Service Installation

Complete the following questions after carefully reading the corresponding chapter.

1. Electrical service equipment consists of the service disconnecting means and the meter _____.

____

2. _____ *True or False?* When the service disconnecting means is the main breaker of the distribution panelboard, the distribution panel is connected with only three conductors.

3. The width of the working space for a residential distribution panel must be a minimum of _____ inches.

____

4. _____ The width of the working space for a distribution panel is required to allow for a _____-degree opening of hinged equipment doors.
   A. 45
   B. 90
   C. 180
   D. There is no *NEC* requirement.

5. A service head, commonly called a(n) _____ head, prevents rain intrusion into the mast and service equipment.

____

6. _____ Above an alley that is accessible to truck traffic, the clearance for overhead service conductors is required to be _____ feet.
   A. 10
   B. 12
   C. 15
   D. 18

7. _____ *True or False?* A service head is *not* required for through-the-roof mast overhead services.

8. _____ The *NEC* requires underground service conductors to be buried at a depth of at least _____.
   A. 15 inches
   B. 30 inches
   C. 3 feet
   D. 6 feet

9. _____ *True or False?* The *NEC* does not allow subpanels to be installed in detached garages.

10. Connecting a conductor of the electrical system to ground through one or more grounding electrodes is called _____ grounding.

____

11. _____ *True or False?* If there is an effective ground-fault current path, fault current on the metal frame of a machine will first pass through the equipment grounding conductor of the machine's cord.

Notes

Name ______________________________ Date ______________ Class __________

CHAPTER 26 Device and Appliance Trim Out

Complete the following questions after carefully reading the corresponding chapter.

1. A leading cause of electrical house fires is improper device _____.

____

2. _____ Which two colors cannot be used to re-identify a white conductor that provides constant power to a switch?
   A. Black and white
   B. Green and red
   C. Red and black
   D. White and green

3. _____ *True or False?* Four-way switches are connected only to the traveler conductors, not the common conductor.

4. The _____ conductor is not connected to a four-way switch.

____

5. If one-half of a receptacle outlet has constant power and the other half is switch-controlled, the outlet is said to be _____.

____

6. _____ *True or False?* It is a best practice to switch the bottom half of a split-wired receptacle and have the top half receive constant power.

7. A GFCI receptacle can protect outlets wired downstream through a connection known as a _____ configuration.

____

8. _____ *True or False?* The line terminations on a GFCI receptacle are connected to the protected downstream outlets.

9. _____ The *NEC* requires the enclosure for a receptacle in wet locations be weatherproof when the receptacle is _____.
   A. in use
   B. not in use
   C. both in use and not in use
   D. There is no *NEC* requirement.

10. Appliances that are served by an individual branch circuit that supplies no other loads are referred to as _____ appliances.

____

11. _____ *True or False?* A circuit breaker can never be used as a disconnecting means for a hard-wired appliance.

12. _____ *True or False?* A disconnect has two terminal lugs for the line connection and two for the load connection.

Notes

Name ______________________ Date ____________ Class ____________

CHAPTER 27 Luminaire Trim Out

Complete the following questions after carefully reading the corresponding chapter.

1. _____ Which type of lamp has the lowest power loss due to heat waste?
   A. Fluorescent
   B. Halogen
   C. Incandescent
   D. LED

2. Recessed can luminaires are composed of two parts: the housing and the _____.

   ____

3. _____ Most of the newer-style canless LED recessed luminaires are _____.
   A. flush-mounted
   B. semi-flush-mounted
   C. Non-Type IC
   D. Type IC

4. _____ The *NEC* requires a minimum of _____ inch(es) of clearance between combustible materials and a Non-Type IC rated luminaire.
   A. 1/2
   B. 1
   C. 3
   D. 6

5. _____ *True or False?* Flush-mounted luminaires are always mounted on ceilings.

6. _____ For track lighting, the *NEC* requires that a single section of track four feet or shorter shall have _____ support(s) unless identified for supports at greater intervals.
   A. one
   B. two
   C. three
   D. four

7. _____ The restricted zone for certain types of luminaires extends to _____ feet horizontally from the bathtub rim.
   A. 3
   B. 5
   C. 8
   D. 10

8. _____ *True or False?* The *NEC* requires one luminaire above each bathroom sink.

9. _____ *True or False?* The installation of partially enclosed fixtures in closets has never been permitted by the *NEC*.

10. A surface-mounted LED luminaire with a completely enclosed light source requires _____ inch(es) of clearance between the luminaire and the closet storage space.

    ____

Notes

Name ______________________ Date ____________ Class __________

CHAPTER 28 Distribution Panel Trim Out

Complete the following questions after carefully reading the corresponding chapter.

1. _____ *Section 408.41* allows _______ neutral conductor(s) per terminal in a panelboard.
   A. one
   B. two
   C. three
   D. four

2. Time current curves show the length of time it takes a circuit breaker to open in relation to the amount of _____ present.

   ____

3. _____ *True or False?* A smaller-size conductor than is required can be terminated on a circuit breaker to accommodate voltage drops on long conductor runs.

4. _____ Using *Table 310.16*, which of the following breakers would be used with a 3 AWG UF wire?
   A. 80
   B. 90
   C. 100
   D. 110

5. _____ The *NEC* limits overcurrent protection for 12 AWG to _______ amps.
   A. 15
   B. 20
   C. 30
   D. 40

6. Using *Table 310.16*, the proper overcurrent protection for a 4 AWG UF wire is _____ amps.

   ____

7. Using *Table 310.16*, the proper overcurrent protection for a 1 AWG NM wire is _____ amps.

   ____

8. _____ *True or False?* Single-pole circuit breakers take up one space in a residential panel.

9. _____ *True or False?* The *NEC* requires GFCI protection for all 120-volt 15- and 20-amp receptacles installed in basements and garages.

10. In non-GFCI-protected breakers, only the _____ conductor of the branch circuit connects to the circuit breaker.

    ____

11. _____ *True or False?* The *NEC* permits the use of dual-purpose breakers to provide GFCI and AFCI protection to kitchen branch circuits.

12. _____ Which of the following is an acceptable way to label a circuit breaker?
    A. Dining room
    B. Guest bedroom
    C. Receptacles
    D. All of the above.

Notes

Name ______________________ Date ____________ Class ________

CHAPTER 29 Branch Circuit Testing and Troubleshooting

Complete the following questions after carefully reading the corresponding chapter.

1. _____ *True or False?* Electrical troubleshooting is performed while the trim-out phase is in progress.
2. Before turning on power to any panelboards, it is best practice to ensure all the terminations are properly _____.

   ____

3. At the panel, voltage from line 1 to the grounded, neutral conductor should be exactly _____ of the line-to-line voltage.

   ____

4. _____ *True or False?* A receptacle tester is the only way to test a receptacle for proper operation.
5. _____ *True or False?* Some receptacle testers can test for proper operation of a GFCI receptacle.
6. _____ *True or False?* Arc-fault breakers can be tested by using a button on the breaker.
7. _____ When a luminaire is connected to a single-pole switch, which of the following *cannot* occur when the switch is turned on?
   A. The circuit breaker protects the circuit trips.
   B. The luminaire turns on.
   C. The luminaire does not turn on.
   D. All of the above can occur.
8. _____ When testing four-way switching circuits, which switches should be tested first?
   A. Single-pole switches
   B. Three-way switches
   C. Four-way switches
   D. The order is not important.
9. Most 240-volt appliances are _____-wired.

   ____

10. _____ *True or False?* Most 120/240-volt circuits are cord-and-plug connected.

Notes

Name ______________________________ Date ______________ Class __________

CHAPTER 30 Outdoor Lights and Receptacles

Complete the following questions after carefully reading the corresponding chapter.

1. _____ The *NEC* requires a minimum conductor size of 10 AWG copper for overhead spans up to _____ feet.
   A. 20
   B. 30
   C. 40
   D. 50

2. If an overhead conductor span crosses over a residential driveway, it must have a clearance of _____ feet.

   ____

3. _____ Underground wiring is used to _____.
   A. prevent contact with electrical conductors
   B. protect conductors from physical damage
   C. Both of the above.
   D. Neither of the above.

4. _____ Direct burial for rigid metal conduit in locations not specified in *Table 300.5(A)* must be at a depth of at least _____ inches.
   A. 6
   B. 12
   C. 18
   D. 24

5. The most popular wire type for residential underground conduit applications is _____ insulation.

   ____

6. Fittings that can withstand exposure to a beating rain are referred to as _____.

   ____

7. _____ Using the formula $E = 4.056 \times 10^{-4} \times l \times \Delta t$, what is the maximum length of conduit that can be used without an expansion fitting if the annual temperature change is 68°F?
   A. 6 feet
   B. 8 feet
   C. 10 feet
   D. 12 feet

8. _____ *True or False?* The *NEC* does permit the splicing of conductors in the pole of a post light if the splices are accessible by removing the luminaire.

9. _____ *True or False?* Extra-duty covers for outdoor receptacles are easily removable.

10. _____ Conduit-supported receptacle outlets must be supported by a minimum of _____ conduits threaded wrench tight into the hubs.
    A. two
    B. three
    C. four
    D. There is no *NEC* requirement.

11. _____ *True or False?* The only acceptable method for supporting freestanding receptacle outlets is with conduit support.

Notes

Name ______________________ Date ______________ Class __________

CHAPTER 31 Swimming Pools and Hot Tubs

Complete the following questions after carefully reading the corresponding chapter.

1. _____ *True or False?* All receptacles must be kept a minimum of 20 feet from the inside wall of a permanently installed pool.

2. The receptacle for a permanently installed pool cannot be located more than _____ feet above the grade level of the pool.

   ____

3. The disconnecting means for all pool-related equipment must be readily accessible and in _____ of the equipment.

   ____

4. _____ *True or False?* The equipotential bonding conductor for a pool must be a solid, not stranded, copper wire.

5. _____ For a new outdoor pool installation, overhead lighting must have a minimum clearance of at least _____ feet above the maximum water level of the pool.
   A. 6
   B. 12
   C. 18
   D. 24

6. A luminaire that is installed in the wall or floor of the pool below the water level, with the luminaire submerged in water, is called a(n) _____ luminaire.

   ____

7. _____ Receptacles for lighting a storable pool must be located at least _____ feet from the inside walls of the pool.
   A. 6
   B. 8
   C. 12
   D. 16

8. _____ *True or False?* The electrocution hazards between the water of a hot tub or spa are the same for swimming pools.

9. _____ Switches for an indoor hot tub must be located a minimum of _____ feet from the inside walls of the tub.
   A. 3
   B. 5
   C. 8
   D. 10

10. _____ *True or False?* No luminaires are permitted to be closer than 7′6″ from the top of a hot tub.

11. _____ *True or False?* If the supply receptacle for a cord-and-plug-connected hydromassage tub is accessible through a service access opening, the receptacle cannot be installed more than one foot from the opening.

12. _____ All metal fittings of a hydromassage bathtub that are in contact with circulating water must be bonded together with a solid copper bonding jumper no smaller than _____ AWG.
    A. 8
    B. 10
    C. 12
    D. 14

Notes

Name ______________________ Date ______________ Class __________

CHAPTER 32 Voice, Data, and Signaling

Complete the following questions after carefully reading the corresponding chapter.

1. An intersystem bonding termination is required to have the capacity for the connection of a minimum of _____ intersystem bonding conductors.

____

2. _____ *True or False?* The *NEC* requires bonding conductors for an IBT to be insulated or covered.

3. _____ The longest a bonding conductor for an IBT can be without having to add an additional ground rod is _____ feet.
   A. 10
   B. 20
   C. 50
   D. 100

4. The most common type of cable used for telephone and networks is _____.

____

5. _____ *True or False?* CAT3 cable can transmit voice signals but is not suitable for data transmission.

6. _____ Which type of cable consists of concentric layers of conductor and insulation?
   A. Coaxial
   B. Fiber optic
   C. Twisted pair
   D. None of the above.

7. _____ Which of the following cables has the highest maximum bandwidth?
   A. CAT3
   B. CAT5e
   C. CAT6A
   D. Coaxial

8. _____ Which of the following conduit types can be used for communication cable runs?
   A. Electrical metallic tubing
   B. Electrical nonmetallic tubing
   C. Flexible metallic conduit
   D. All of the above.

9. _____ *True or False?* Orange is a common color choice for electrical nonmetallic tubing that is being used for communication and signaling circuits.

10. The bend radius for two-inch ENT is _____ inches.

____

11. _____ The least expensive and easiest to install connector for coaxial cable is the _____ connector.
   A. compression
   B. crimped
   C. twist-on
   D. They all have similar difficulty and price.

12. _____ *True or False?* Testers for fiber optic cable must use a light signal to test the quality of the connections.

13. _____ When mounting a smoke detector on a wall, the detector should be no more than _____ inches from the ceiling.
A. 2
B. 8
C. 12
D. 20

Name ______________________ Date ______________ Class ____________

CHAPTER 33 Solar Photovoltaic Systems

Complete the following questions after carefully reading the corresponding chapter.

1. During site assessment for PV power systems, a(n) _____ analysis is performed to determine if the system will be blocked from the sun by large trees or other objects.

 ____

2. _____ *True or False?* The amount of electrical current produced by a solar cell is 0.5 amps.

3. _____ The *NEC* requires that PV output circuits be sized based on _____ percent of the output rating of the parallel connected inverters.
   A. 75
   B. 100
   C. 125
   D. 150

4. _____ *True or False?* String inverters *cannot* be used in an off-grid system.

5. The *NEC* requires that PV systems connected to the utility have _____ technology to turn the inverter system off when ac power is lost.

 ____

6. An arrangement that allows homes with battery storage to provide power independent of the utility is commonly called _____.

 ____

7. _____ *True or false?* The racking for PV systems provides better support if alternate framing members are used.

8. _____ Why do installers often place screening around a PV array?
   A. To prevent animals from getting under the array.
   B. To prevent snow from getting under the array.
   C. To prevent sunlight from getting under the array.
   D. To prevent water from getting under the array.

9. _____ *True or False?* Grounding is not required for pole-mounted PV arrays.

10. The electrical wiring, devices, equipment, conduits, and cables for a PV array are called the _____ components.

 ____

11. _____ *True or False?* All utilities and municipal codes require a PV system disconnect to disconnect the photovoltaic production equipment and the energy storage equipment.

12. _____ The overcurrent devices that supply power from a PV system need to be _____.
   A. located inside the dwelling
   B. rated at 200 amps or greater
   C. rated for a continuous load
   D. All of the above.

Notes

Name ______________________________ Date ______________ Class __________

CHAPTER 34 Garage Wiring and Electrical Vehicle Charging Power

Complete the following questions after carefully reading the corresponding chapter.

1. _____ *True or False? NEC* requirements for a detached garage require two outlets for each vehicle stall.

2. The *NEC* requires detached garage receptacles be installed no higher than _____ feet above the floor.

   ____

3. _____ If a garage door opener is connected to a circuit with lighting, the garage door opener rating cannot exceed _____ percent of the circuit rating.
   A. 25
   B. 50
   C. 75
   D. 90

4. _____ Running power to a separate structure requires _____.
   A. additional grounding
   B. a separate disconnecting means
   C. Both of the above.
   D. None of the above.

5. _____ *True or False?* Electrical boxes should *not* be installed within the same vertical stud space on both sides of a fire-rated wall without addition fire protection.

6. Receptacles in an attached garage must be supplied by a(n) _____-amp branch circuit.

   ____

7. _____ *True or False?* The *NEC* does permit lighting for general illumination of electrical equipment in an attached garage to be controlled by an automatic means.

8. Circuits used to charge electric vehicles operate as _____ loads, which are electrical loads where the maximum current is expected to continue for three hours or more.

   ____

9. _____ With portable cord-and-plug equipment for an EVSE, the maximum length of the supply cord to the receptacle is _____ inches.
   A. 6
   B. 12
   C. 24
   D. 36

10. _____ *True or False?* An EVSE is required to be mounted at a height where the cord does not touch the floor.

11. _____ A 240-volt Level II type EVSE with an amp rating of 30 amps is to be installed in a residential garage. What is the minimum overcurrent device required to supply this EVSE?
    A. 20-amp
    B. 30-amp
    C. 40-amp
    D. 50-amp

12. Receptacles supplied from a feeder and subpanel in a garage are to be calculated at _____ VA per device.

    ____

13. _____ If two luminaires are installed in a garage, each having a rating of 100 watts, but there is only one 10-watt LED lamp installed, the total feeder calculation for the circuit would be _____ watts.
A. 10
B. 20
C. 100
D. 200

Name ______________________________ Date ______________ Class __________

CHAPTER 35 Standby Power Systems

Complete the following questions after carefully reading the corresponding chapter.

1. _____ *True or False?* Residential standby power systems are intended for life-saving power options.

2. Loads that use magnetic fields that can store energy and return some of the stored energy back to the source are called _____ loads.

   ____

3. _____ Which of the following loads has the largest running load? Refer to Figure 35-1.
   A. Coffee maker
   B. Dishwasher
   C. Microwave
   D. Space heater

4. _____ Which of the following loads has the largest surge load? Refer to Figure 35-1.
   A. Clothes washer
   B. Dehumidifier
   C. Electric range
   D. Garage door opener

5. _____ *True or False?* Standby systems that power the entire home require a main breaker ahead of the main electrical panel.

6. Standby power systems are able to use renewable energy sources using _____ interconnect devices.

   ____

7. _____ The *NEC* requires that all serviceable equipment installed around the front of a permanent generator allow a minimum of _____ inches of working space.
   A. 18
   B. 24
   C. 30
   D. 36

8. A generator rated at 24 kW has a maximum circuit breaker size of _____ amps.

   ____

9. _____ The minimum size for a copper equipment grounding conductor for a generator rated at 26 kW is _____ AWG.
   A. 2
   B. 4
   C. 6
   D. 8

10. _____ *True or False?* All installations of wiring for standby power using a portable generator require some means to isolate the utility connection.

11. _____ *True or False?* An advantage of battery-powered backup is that no consideration for dissipating heat is necessary when selecting a location for the cabinet.

Notes

Name ______________________ Date ____________ Class ____________

CHAPTER 36 Troubleshooting Basics

Complete the following questions after carefully reading the corresponding chapter.

1. _____ *True or False?* With an overload condition, the current stays in the conductive pathway of the circuit.

2. _____ Which of the following is the first sign of an overload condition?
   A. GFCI breaker opening
   B. No current flowing
   C. Smell of heated insulation
   D. All of there above.

3. The metal enclosed panelboard that contains the circuit breakers that supply a home's power and lighting circuits is called the _____ center.

   ____

4. _____ *True or False?* The half-split method is a technique used to divert half of the current from an overloaded circuit to a second branch circuit.

5. _____ In a multi-family dwelling unit fed by a three-phase systems, each dwelling unit is fed by a 3-wire single-phase system with a voltage between the ungrounded conductors of _____ volts.
   A. 120
   B. 180
   C. 208
   D. 240

6. A short circuit that includes the equipment grounding conductor in the short circuit current pathway is called a(n) _____.

   ____

7. Connecting two ungrounded conductors of opposite phases together will cause a(n) _____ circuit.

   ____

8. _____ The first step in troubleshooting a ground fault is to _____.
   A. connect an ohmmeter to the wiring to check for continuity
   B. connect a voltmeter to the wiring to check for continuity
   C. remove all loads from the circuit
   D. reset the circuit breaker

9. _____ When searching for a ground fault, a(n) _____ can be used to check for continuity.
   A. ammeter
   B. ohmmeter
   C. voltmeter
   D. None of the above.

10. _____ *True or False?* The half-split troubleshooting method *cannot* be used to isolate the source of a short circuit.

11. _____ To protect equipment from damage, NEMA recommends a tolerance of no more than _____ percent of the nominal rated voltage.
    A. 1
    B. 2
    C. 5
    D. 10

12. _____ *True or False?* Opening a neutral can create a condition resulting in undervoltage of equipment with high resistances.

13. _____ *True or False?* Loose connections in a circuit can decrease the resistance of a circuit.

14. _____ *True or False?* In the search for an arc-fault, an unintentional connection between the grounded conductor and the equipment grounding conductor of the circuit be found using a continuity tester.

Name ______________________ Date ____________ Class ____________

CHAPTER 37 Removing and Adding Devices and Luminaires

Complete the following questions after carefully reading the corresponding chapter.

1. _____ Receptacles designed for 250-volt loads should only be used to connect loads that are designed to operate _____.
   A. at any voltage from 0 to 250 volts
   B. at more than 120 volts but less than 250 volts
   C. at 250 volts
   D. There are no voltage restrictions.

2. _____ *True or False?* The current rating of a single receptacle can be less than the current rating of the individual branch circuit as long as the receptacle's current rating is greater than that of the loads connected.

3. _____ The three-wire circuit for existing ranges and dryers does not contain _____.
   A. a grounded conductor
   B. an equipment grounding conductor
   C. one of the hot, ungrounded conductors
   D. None of the above.

4. _____ *True or False?* Split-circuit receptacle installations are common in new construction.

5. _____ In a split-circuit installation, if the receptacle is part of a multi-wire branch circuit, _____.
   A. the tab between the terminals for the grounded conductor is removed
   B. the tab between the terminals for the ungrounded conductor is removed
   C. both the grounded and ungrounded tabs are removed
   D. No tabs need to be removed.

6. For a switched receptacle installation, remove the tab between the _____ conductor terminals from the receptacle.

   ____

7. _____ *True or False?* When replacing receptacles in existing installations, if the existing receptacle does not have GFCI protection, but newer *NEC* regulations require GFCI protection, then the replacement receptacle must have the protection.

8. _____ *True or False?* Some manufacturers require that their appliances be fed by a dedicated branch circuit.

9. _____ An advantage of placing a new outlet on an interior wall is that interior walls are _____.
   A. closer to the panelboard
   B. less likely to have insulation
   C. never load bearing
   D. All of the above.

10. When cutting into sheetrock, most electricians use a(n) _____ saw to do the job.

    ____

11. Nonmetallic boxes are widely used, but they are not allowed to be used for _____ cabling methods.

    ____

12. _____ *True or False?* CFL technology has the longest lamp life.

13. A luminaire that focuses 90 percent or more of the light produced in the downward direction is considered _____ lighting.

    ____

Notes

Name ______________________________ Date ______________ Class __________

CHAPTER 38 Adding a Circuit

Complete the following questions after carefully reading the corresponding chapter.

1. _____ Fish tapes are commonly made out of _____.
   A. fiberglass
   B. nylon
   C. steel
   D. All of the above.

2. _____ *True or False?* Fish sticks are more rigid and less flexible than fish tape.

3. A vertical void in building construction that allows for the installation of pipes, wire, and heating and cooling through a building is called a(n) _____.

   ____

4. _____ *True or False?* The *NEC* permits installing nonmetallic sheathed cable in return air cavities only if the cable is run parallel to the long dimension of the cavity.

5. _____ The *NEC* requires nonmetallic sheathed cables installed in open attic spaces to be supported at intervals no greater than _____ feet.
   A. 3
   B. 4-1/2
   C. 6
   D. 7-1/2

6. _____ *True or False?* Knockouts are designed to be pushed out of the enclosure from the inside.

7. When fishing cables into panelboards that have the main breaker, removing the utility _____ is often the only way to de-energize the panelboard.

   ____

8. _____ On all new construction, the *NEC* requires the installation of an emergency disconnecting means on the _____ of single-family dwellings.
   A. exterior
   B. interior
   C. Both A and B.
   D. No emergency disconnecting means is required.

9. Twin breaker, duplex breaker, and double breaker are all common names for a(n) _____ breaker.

   ____

10. If a 15-space, 15-circuit panelboard contains 10 standard breakers, _____ tandem breakers can be added before the panelboard is in violation of the rating.

    ____

11. A shared grounded conductor, which could exceed the ampacity of the conductor, is a potential problem when tandem breakers are installed in systems using _____ branch circuits.

    ____

12. _____ *True or False?* A 60-amp subpanel can be connected to a 100-amp feeder.

13. _____ Which of the following locations is an acceptable place to install a subpanel?

A. Bathroom
B. Clothes closet
C. Over stairs
D. Unfinished basement

Name ______________________________ Date ______________ Class __________

CHAPTER 39 Service Upgrades

Complete the following questions after carefully reading the corresponding chapter.

1. One common reason to upgrade electrical service is to replace _____ with circuit breakers.

____

2. _____ Which of the following can affect whether a new electrical service must be moved to a different location than the one it is replacing?
   A. Changes to the *NEC*
   B. Addition of a new door
   C. Installation of a deck
   D. All of the above.

3. _____ *True or False?* Installing a service disconnect on the exterior of the building allows for unlimited length feeder conductors to enter the dwelling.

4. _____ *True or False?* Type SE cable is required to be contained with a raceway.

5. _____ *True or False?* Metallic raceways are more affected by thermal contraction and expansion than PVC conduit.

6. _____ An advantage of choosing a copper service conductor over aluminum is that _____.
   A. copper performs better in areas exposed to moisture
   B. copper is more flexible
   C. copper is more readily available
   D. None of the above.

7. The *NEC* requires the mitigation of the buildup of condensation by _____ conduits that travel between areas of major temperature difference.

____

8. Grounding electrode conductors are based on the size of the _____ service entrance conductors.

____

9. _____ If the size of the largest ungrounded conductor in a service entrance is 2/0 copper, what size copper-clad aluminum grounding conductor is required?
   A. 6
   B. 4
   C. 2
   D. 1/0

10. In areas where inspections are not readily available, a utility might allow an installer to complete a(n) _____, a written statement affirming that an installation is compliant with the *NEC*.

____

11. _____ *True or False?* The *NEC* suggests but does not require a panelboard to be labeled in a clear and legible manner.

Notes

Name ______________________ Date ____________ Class __________

Lab Activity 1A

Exploring Employment Opportunities

Objectives

After completing this lab activity, you will be able to:

- Use job-search websites to find employment opportunities.

Text Reference

Chapter 1, *Getting Started in the Industry*

Materials/Equipment

The following equipment is needed to perform these activities:

- Computer or device with internet access

Introduction

Discovering employment opportunities is critical to your professional success. At any time, there may be several job openings available to you that may provide higher wages, better working conditions, and more opportunities for advancement. It is up to you to look for better professional opportunities.

You can learn about employment opportunities in many ways. Your school may have a career center or job board to help connect students with employers. Your instructors may know of employers searching for entry-level electricians. Many companies post job openings on their websites. Job-search websites provide an easy way to see job posting from many companies. As you begin to work in the electrical industry, you can learn about employment opportunities through networking—talking with your coworkers and people you meet from other companies.

Using a Job-Search Website

Use a job-search engine such as Indeed, Monster, or CareerBuilder to complete this activity.

1. Find a job posting for an entry-level electrical position in your area. If job postings in your immediate area are limited, expand your search until you find a job posting.
2. Summarize the job posting, including the following (if provided):
   - Company

     ____
   - Position

     ____
   - Salary

     ____

- Responsibilities of the position (3–5)

- Requirements for applicants

3. What about this job posting do you find interesting?

4. Find a job posting for a more advanced (3–10 years of experience) electrical-related position in your area.
5. Summarize the job posting, including the items listed in the earlier question.

6. What about this job posting do you find interesting?

Name ___________________________________

7. Select another city in the United States. This could be a city you are interested in visiting. Find a job posting for either an entry-level or more advanced electrical-related position in that city.
8. Summarize the job posting, including the items listed in the earlier question.

____

____

____

____

____

____

____

9. What about this job posting do you find interesting?

____

____

____

____

____

Notes

Name ______________________ Date ____________ Class ________

Lab Activity 1B

Preparing a Résumé

Objectives

After completing this lab activity, you will be able to:

- Prepare a résumé.

Text Reference

Chapter 1, *Getting Started in the Industry*

Materials/Equipment

The following equipment is needed to perform these activities:

- Computer with word-processing software and internet access

Introduction

A résumé is a brief outline of your education, work experience, and other qualifications for employment. A well-written and thoughtful résumé can help you get an interview. You will need to include several sections on your résumé. Textbook Figure 1-1 is an example of a résumé you can use to create your own résumé. Consider the following when preparing your résumé:

- For items in the *Work Experience* section, list several key responsibilities of the position. Be sure to include responsibilities related to interacting with coworkers (such as training new employees) and customers.
- List work experience in reverse chronological order, beginning with the most recent.
- If you have many positions to list in the *Work Experience* section, you can rename the section as *Recent Work Experience* or *Related Work Experience*. In these cases, you would list only your most recent positions or the positions most closely related to the position for which you are applying. You can also add an *Other Work Experience* item and list some additional positions without responsibilities.
- In addition to the *Education* and *Work Experience* sections, you can include a *Skills* section. You can list electrical skills, computer skills, communication skills, and other skills in this section.
- Never include untrue information on your résumé. If an employer hires you and then discovers that information on your résumé was inaccurate, the employer may terminate your employment.

Preparing a Résumé

1. Using word-processing software such as Microsoft Word or Google Docs, prepare your résumé. You can use the sample in textbook Figure 1-1 as a model, and you can also search online for résumé templates.
2. Ask one or two people to read your résumé and provide suggestions for improving it. Preferably, choose reviewers with some work experience or experience with résumés.
3. Revise your résumé based on the suggestions from your reviewers.

Notes

Name ______________________ Date ______________ Class __________

Lab Activity 1C

Preparing for an Employment Interview

Objectives

After completing this lab activity, you will be able to:

- Prepare for an employment interview.

Text Reference

Chapter 1, *Getting Started in the Industry*

Materials/Equipment

The following equipment is needed to perform these activities:

- Computer or device with internet access

Introduction

Walking into a job interview unprepared will decrease your chances of receiving an offer of employment. A candidate who has prepared prior to an interview is likely to stand out. In this activity, you will perform some steps to help you prepare for an interview.

- Have a tidy appearance for your interview and dress appropriately.
- Research the company prior to the interview and look for opportunities to mention what you know about the company during the interview.
- Ask questions about the position and company during the interview.
- Be on time for your interview.

Preparing for an Employment Interview

Interview preparation can be separated into two categories: general preparation and interview-specific preparation. General preparation focuses on you and your résumé and does not change from interview to interview. Interview-specific preparation focuses on the specific position and company and is different for each interview.

General Interview Preparation

1. Review your résumé. Identify five specific examples of situations that you may want to mention in your interview. These situations may be related to your work experience, your experience in school, or an experience from your personal life. You will be able to use these examples when responding to interview questions. Some possible situations to consider include the following:
   - Examples of times you needed to learn a new skill. How did you learn the skill? How were you able to apply that new skill in the future?
   - Examples of challenges that you overcame. What was the challenge? How were you able to overcome it? How did overcoming that challenge benefit you in the future?
   - Examples of disagreements with coworkers. How do you resolve the disagreement?
   - Examples of interactions with an unhappy customer. How did you address the customer's concerns? What was the final resolution?
   - Examples of good performance, typically in school or at work. How did you know your performance was good?
   - Examples of working as part of team. What was your role? How do you interact with other team members? What goal did the team accomplish?

List your five examples below. You do not need to describe them in great detail. A simple sentence or two summarizing the example is sufficient.

Example 1:

Example 2:

Example 3:

Example 4:

Example 5:

Name ____________________

2. Asking good questions in an interview can help you learn more about the position and the company, while also helping you stand out as a potential candidate. Search online for "good questions for an interview candidate to ask" or similar. Review a few of the web pages and articles you find, and then identify five potential questions you may want to ask a person who is interviewing you. List the questions below.

   Question 1:

   Question 2:

   Question 3:

   Question 4:

   Question 5:

Interview-Specific Preparation

Select a job posting to use to answer the following questions. You can select one of the job postings you identified in an earlier lab activity, or you can select a job posting from a job-search website or other source.

1. What is the name of the company offering the position?

____

2. What is the position listed in the posting?

____

3. Go to the company's website and record five facts about the company below. Try to identify the mission and/or values of the company. Often, these can be found on an "About Us" web page or as part of the job posting. During an interview, you will be able to explain how the company's mission or values are aligned with your personal ideals.

   Company fact 1:

   ____

   ____

   Company fact 2:

   ____

   ____

   Company fact 3:

   ____

   ____

   Company fact 4:

   ____

   ____

   Company fact 5:

   ____

   ____

4. Review the information provided about the position and write three related questions you could ask during an interview.

   Question 1:

   ____

   ____

   ____

   Question 2:

   ____

   ____

   ____

   Question 3:

   ____

   ____

   ____

Name ______________________ Date ____________ Class ________

Lab Activity 1D

Introduction to the *National Electrical Code*

Objectives

After completing this lab activity, you will be able to:

- Identify methods of finding information in the *National Electrical Code*.

Text Reference

Chapter 1, *Getting Started in the Industry*

Materials/Equipment

The following equipment is needed to perform these activities:

- Copy of the *National Electrical Code*

Introduction

The *National Electrical Code* (*NEC*) provides guidelines and requirements to ensure a safe electrical installation. Electricians must be familiar with the requirements applicable to the work being performed. In addition, electricians must be comfortable navigating the *NEC* to find specific requirements. This lab activity provides an introduction to begin your familiarization with the *NEC*.

Introduction to the *National Electrical Code*

Use a copy of the 2023 edition of the *National Electrical Code* to answer the following questions.

1. Refer to *Figure 90.3* in *Article 90*. Which chapters of the *NEC* generally apply to all electrical installations?

   ____

2. Use the *NEC* index to answer the following question: Which *NEC* section contains information about grounding electrode conductors for solar photovoltaic systems?

   ____

3. Use the *NEC* index to answer the following question: Which *NEC* section contains information about cable fill in cable trays?

   ____

4. Use the *NEC* index to answer the following question: Which *NEC* section contains information about insulation for service-entrance conductors?

   ____

5. Definitions are found in *NEC Article 100*. According to the definition of "kitchen," what must be present for an area to be considered a kitchen?

   ____

   ____

   ____

6. According to their definitions, how does a "rainproof" device differ from a "raintight" device?

7. According to its definition, what two items do "service conductors" connect?

8. Refer to the contents in the *NEC* and list five examples of special occupancies addressed in *Chapter 5*.

9. Refer to the contents in the *NEC* and list five examples of special equipment addressed in *Chapter 6*.

10. Refer to the contents in the *NEC* and list five examples of special conditions addressed in *Chapter 7*.

Name ______________________________ Date ______________ Class __________

Lab Activity 2A

Lockout/Tagout

Objectives

After completing this lab activity, you will be able to:

- Perform a lockout/tagout (LOTO) procedure.

Text Reference

Chapter 2, *Safety in Residential Wiring*

Materials/Equipment

The following tools and equipment are needed to perform these activities:

- Electrical system with a panel and a receptacle
- Noncontact voltage tester or receptacle tester
- Breaker lock lockout devices

Introduction

Lockout/tagout procedures help to ensure the safety of an electrician by preventing a circuit being worked on from becoming energized. Lockout/tagout procedures are critical in industrial and commercial settings, where many workers may have access to a power panel.

Safety Review

Follow all safety procedures specified by your instructor and in any appropriate manufacturer's information. The following are some additional safety-related items to keep in mind:

- Be very cautious when taking measurements on energized systems.
- Do not contact any exposed conductors or terminals.

Locking Out a Circuit

The following is a general procedure. Your instructor may modify this procedure based on the equipment and instruments available in the lab.

1. Using the noncontact voltage tester or receptacle tester, confirm that voltage is present at the receptacle.
2. At the panelboard, switch the circuit breaker for the receptacle's circuit to the OFF position.
3. Place a breaker lock lockout device on the circuit breaker.
4. At the receptacle, use the noncontact voltage tester or receptacle tester to confirm that no voltage is present.
5. Return to the panelboard, remove the breaker lock, and switch the circuit breaker to the ON position.

Reflect and Analyze

1. What is the benefit of locking out a circuit breaker compared with simply switching the circuit breaker to the *off* position?

2. Why do you test for voltage at the device after locking out the circuit breaker?

3. You switch a circuit breaker to the *off* position and install a circuit breaker lock. When you return to the device and check for voltage, you find that the circuit is still energized. How could this happen?

Name ______________________ Date ____________ Class ____________

Lab Activity 2B

Personal Protective Equipment (PPE) Inspection and Fit Test

Objectives

After completing this lab activity, you will be able to:

- Inspect personal protective equipment (PPE).
- Test PPE for proper fit.

Text Reference

Chapter 2, *Safety in Residential Wiring*

Materials/Equipment

The following tools and equipment are needed to perform these activities:

- Safety glasses
- Hard hat

Introduction

You cannot predict when a hazardous event will occur. Therefore, it is critical to always wear appropriate personal protective equipment (PPE) when you are on a jobsite. If the PPE does not fit correctly or is damaged in some way, it may not provide the required protection. Developing the habit of inspecting your PPE and ensuring a proper fit will help ensure your safety throughout your career.

Safety Glasses Inspection and Fit Test

1. Do your safety glasses include a Z87+ mark?

   ____

2. Careful inspect your safety glasses. Are there any chips or cracks in the protective surface? Are the arms (temples) in good condition?

   ____

   ____

3. If the arms (temples) on the safety glasses are adjustable, check that the adjustment mechanism functions correctly.
4. Wear the safety glasses. Do they fit snuggly? If you wear prescription glasses, are the safety glasses large enough to provide protection over your prescription glasses?

   ____

   ____

5. Nod and shake your head quickly. Did the safety glasses move? If so, adjust the arms (temples) and test again.

   ____

   ____

Hard Hat Inspection and Fit Test

1. Find the certification label on your hard hat. Record the ANSI standard, type, and class information listed on the label.

2. Inspect the shell of the hard hat. Are there any visible chips or cracks in the shell?

3. The assembly of straps inside the shell is called the suspension. Carefully inspect all of the straps in the suspension. Are any straps worn or damaged?

4. The strap at the rear of the hard hat is adjustable. Adjust the strap to ensure that the mechanism functions correctly.

5. Place the hard hat on your head. Nod and shake your head quickly. The hard hat should not move on your head. If the hard hat slid on your head, tighten the adjustable strap and recheck. Continue to adjust the hard hat until a proper fit is achieved.

Reflect and Analyze

1. Why could using a cracked hard hat or safety glasses be dangerous?

2. According to ANSI's hard hat standard, what are the different "type" categories for hard hats, and what is the requirement for each type category?

3. According to ANSI's hard hat standard, what are the different "class" categories for hard hats, and what is the requirement for each class category?

Name ______________________________ Date ______________ Class __________

Lab Activity 3A

Cutting Cable and Stripping Conductors

Objectives

After completing this lab activity, you will be able to:

- Measure and cut cable.
- Strip conductors.

Text Reference

Chapter 3, *Hand Tools*

Materials/Equipment

The following tools and equipment are needed to perform these activities:

- 14/2 or 12/2 NM cable (54″ minimum)
- Tape measure
- Marker or other marking tool
- Lineman (side cutter) pliers
- Diagonal-cutting pliers
- Electrician's knife
- Cable ripper
- Wire stripper or multipurpose tool

Introduction

Cutting cable and stripping conductors are fundamental tasks for an electrician. In this activity, you will practice cutting and stripping. These operations are not difficult, but executing them accurately and correctly is critical to a safe installation.

Safety Review

Follow all safety procedures specified by your instructor and in any appropriate manufacturer's information. The following are some additional safety-related items to keep in mind:

- Wear safety glasses when using any tool.
- Handle tools careful to help avoid accidents and damage to the tool.
- Before using any tool, inspect the tool to ensure it is working properly and is not damaged in any way.

Measuring and Cutting Cable

1. The cable sheathing includes a marking that identifies the size of the conductors. Record the conductor size:
   A. Conductor size: _____ AWG
2. Use the tape measure and place a mark on the cable 18″ from one end.
3. Using the lineman (side cutter) pliers, cut the cable at this mark. Cut the cable perpendicular to the length of the cable, so the cut end is square.
4. Use the tape measure to mark a second 18″ length.
5. Use the diagonal-cutting pliers to cut the cable at the mark. You should now have two 18″ lengths of cable.
6. Use the tape measure to mark a third 18″ length.

7. Use the diagonal-cutting pliers to cut the cable at the mark. You should now have three 18″ lengths of cable.
8. On each of your 18″ lengths of cable, use the tape measure to measure and mark 2″ from one end.
9. Use the lineman pliers to cut each of the three cables at the mark. You should now have three 2″ lengths of cable and three 16″ lengths of cable.

Removing Sheathing

1. Use the tape measure to mark 8″ from an end on each of your 16″ lengths of cable.
2. Use the electrician's knife to remove the sheathing from the 8″ of cable.
   A. Be careful when scoring and cutting the sheathing to not nick the conductor insulation or bare grounding conductor.
   B. Cut around the sheathing at the mark first, and then cut the sheathing along the length of the cable.
   C. Use the electrician's knife to cut away the sheathing and any paper wrapping.
3. Carefully inspect the conductor insulation. Are there any nicks or damage to the insulation?

   ____

4. Repeat this process on the second length of cable, with the following adjustments:
   A. Use the cable ripper when removing the sheathing.
   B. Use the lineman pliers to cut away the sheathing and paper wrapping.
5. Repeat this process on the third length of cables, using whichever tools you prefer. When finished, you should have three lengths of cable with 8″ of exposed conductors on each.

Stripping Conductors

1. Use a wire stripper or multipurpose tool to strip the insulation from 3/4″ of the black and white conductor in each of your three cables.
   A. Place the conductor in the correct stripping slot in the tool based on the conductor size.
   B. Squeeze the tool to cut through the insulation, then slide the tool off the end of the conductor to remove the insulation.
2. Carefully inspect the exposed conductor. Are there any nicks or damage to the conductor?

   ____

Reflect and Analyze

1. Inspect the three 2″ pieces of cable you cut. Are the ends of the cable square? Are there any differences between the pieces? If so, what is different?

   ____
   ____
   ____
   ____
   ____

Name ______________________________

2. When cutting the cable, what differences did you notice (such as required effort, accuracy, etc.) between using lineman pliers and using diagonal-cutting pliers?

3. When removing the sheathing, did you notice any differences between using the cable ripper compared with using the electrician's knife?

4. Did you notice any differences between using the lineman pliers compared with using the electrician's knife to cut away sheathing and paper wrapping?

Notes

Name ______________________ Date ______________ Class ____________

Lab Activity 3B

Using a Tape Measure

Objectives

After completing this lab activity, you will be able to:

- Use a tape measure to measure lengths and distances.

Text Reference

Chapter 3, *Hand Tools*

Materials/Equipment

The following tool is needed to perform these activities:

- Tape measure

Introduction

The tape measure is one of the most common tools for any construction trade. Being able to use it to accurately measure distances is a fundamental skill. The following are some basic tips related to tape measures and their uses:

- Tape measures are available in different lengths and with different blade widths. A 25′ or 30′ tape measure is sufficient for most needs. A 1″ blade will allow the tape to extended farther without bending. A 1/2″ blade may bend too easily, making the tape measure more difficult to use.
- The metal hook at the end of the blade slides slightly in the direction of measurement. The amount of slide is equal to the thickness of the hook. When the hook is positioned on the end of the distance being measured, it slides out so the measurement begins on the inside face of the hook. When the hook is placed against the item being measured, the hook slides in so the measurement begins at the outside face of the hook.
- When retracting the blade into the tape measure, stop the retraction so the blade is still extended a few inches, and then retract the last few inches. If you retract relatively long extensions of the blade without stopping the retraction, the hook will slam into the tape measure housing with a greater amount of force. Over time, this force may damage the slide mechanism for the hook, resulting in inaccurate measurements.

Safety Review

Follow all safety procedures specified by your instructor and specified in any appropriate manufacturer's information. The following are some additional safety-related items to keep in mind:

- When retracting the blade on a tape measure, keep your fingers clear of the blade to avoid possible scrapes, friction burns, or cuts.

Using a Tape Measure

For this activity, your instructor will identify ten items or distances to measure in your classroom and lab. For each item, list the item being measured and record the measurement.

1. Measurement #1

   A. Description of measurement: ______________________________

   B. Measurement: ______________________________

2. Measurement #2
    A. Description of measurement: ______________________________
    B. Measurement: ______________________________
3. Measurement #3
    A. Description of measurement: ______________________________
    B. Measurement: ______________________________
4. Measurement #4
    A. Description of measurement: ______________________________
    B. Measurement: ______________________________
5. Measurement #5
    A. Description of measurement: ______________________________
    B. Measurement: ______________________________
6. Measurement #6
    A. Description of measurement: ______________________________
    B. Measurement: ______________________________
7. Measurement #7
    A. Description of measurement: ______________________________
    B. Measurement: ______________________________
8. Measurement #8
    A. Description of measurement: ______________________________
    B. Measurement: ______________________________
9. Measurement #9
    A. Description of measurement: ______________________________
    B. Measurement: ______________________________
10. Measurement #10
    A. Description of measurement: ______________________________
    B. Measurement: ______________________________

Reflect and Analyze

1. Which measurement was the most difficult to execute? Why?

____

____

____

____

____

Name ____

2. What was your level of experience and comfort using a tape measure prior to completing this activity?

____

____

____

____

3. What is your level of experience and comfort using a tape measure after completing this activity?

____

____

____

____

Notes

Name ______________________ Date ______________ Class __________

Lab Activity 4A

Using a Reciprocating Saw

Objectives

After completing this lab activity, you will be able to:

- Cut wood and metal using a reciprocating saw.

Text Reference

Chapter 4, *Power Tools*

Materials/Equipment

The following tools and equipment are needed to perform these activities:

- Reciprocating saw
- Piece of 2×4 lumber (minimum 16″ long)
- Electrical nonmetal tubing (ENT) (minimum 16″ long)
- Tape measure
- Marker or other marking instrument

Introduction

Reciprocating saws are used in demolition work. Other types of saws are better for making precise cuts, but the reciprocating saw can make quick cuts and is useful in tight spaces.

Various types of blades are used in a reciprocating saw. The blade type is selected based on the type of material being cut. The blades vary in the type of material used to make the blade and the number and size of teeth on the blade.

In most reciprocating saws, the speed of the blade is controlled by the trigger. For most costs, use the highest speed with the trigger pulled to the maximum position.

Safety Review

Follow all safety procedures specified by your instructor and in any appropriate manufacturer's information. The following are some additional safety-related items to keep in mind:

- Wear safety glasses and gloves when using a reciprocating saw.
- When using a reciprocating saw, firmly grip the tool with both hands while making a cut.
- Before using any power tool, review the tool and the user's manual to make sure you understand all the controls and adjustments on the tool.

Cutting Wood with a Reciprocating Saw

1. Secure the piece of 2×4 lumber to a workbench or table using a vise and clamps. Ensure that the piece is secure by pushing and pulling it in different directions to make sure it does not move.
2. Use the tape measure to measure a mark 3″ from the end of the lumber. Mark the lumber across its top and along the side you will cutting from.
3. Detach the battery from the reciprocating saw (if cordless) or ensure that the reciprocating saw is not plugged into a receptacle (if corded).
4. Install an appropriate blade for cutting wood into the reciprocating saw. Use the shortest blade that will extend beyond the material to be cut. Wiggle and pull gently on the blade to ensure that it is installed securely. If you are unsure, check with your instructor.

5. Inspect the reciprocating saw to ensure that it is not damaged in any way. Examine the trigger to see if the saw has a trigger-locking mechanism. If it does, with the saw still disconnected from battery/power pull on the trigger, activate the trigger-locking mechanism, and then release the trigger-locking mechanism. Repeat this until you are comfortable with the activation and deactivation.
6. Check that the trigger is not locked in the ON position, then reattach the battery or plug the power cord into a receptacle.
7. Visualize making the cut through the wood and ensure that there are no obstructions that could interfere with the blade while making the cut.
8. Hold the reciprocating saw firmly in both hands and make the cut.
9. After completing the first cut, use the tape measure to mark a second cut 3″ from end you just cut.
10. Recheck for any obstructions at this cut point. Adjust the position of the wood if needed. If you adjust the position, make sure the wood is secure in its new position.
11. Hold the reciprocating saw firmly in both hands and make the second cut.
12. Unplug the reciprocating saw (if corded) or detach the battery (if cordless).

Cutting Metal with a Reciprocating Saw

1. Secure the piece of ENT to a workbench or table using a vise and clamps. Ensure that the piece is secure by pushing and pulling it in different directions to make sure it does not move.
2. Use the tape measure to measure a mark 3″ from the end of the tubing.
3. Detach the battery from the reciprocating saw (if cordless) or ensure that the reciprocating saw is not plugged into a receptacle (if corded).
4. Install an appropriate blade for cutting metal into the reciprocating saw. Use the shortest blade that will extend beyond the material to be cut. Wiggle and pull gently on the blade to ensure that it is installed securely. If you are unsure, check with your instructor.
5. Check that the trigger is not locked in the ON position, then reattach the battery or plug the power cord into a receptacle.
6. Visualize making the cut through the tubing and ensure that there are no obstructions that could interfere with the blade while making the cut.
7. Hold the reciprocating saw firmly in both hands and make the cut.

SAFETY NOTE

The cut surface of the ENT may have sharp edges and burrs. Sand or deburr the edges on both pieces to remove any sharp bits.

8. After completing the first cut, use the tape measure to mark a second cut 3″ from the end you just cut.
9. Recheck for any obstructions at this cut point. Adjust the position of the tubing if needed. If you adjust the position, make sure the tubing is secure in its new position.
10. Hold the reciprocating saw firmly in both hands and make the second cut.
11. Unplug the reciprocating saw (if corded) or detach the battery (if cordless).

Name ____________________

Reflect and Analyze

1. Examine the cut edges of the lumber pieces and evaluate the location and squareness of your cuts.

2. Examine the cut edges of the ENT pieces and evaluate the location and squareness of your cuts.

3. How did your first cut compare to your second cut? Did you make any adjustments to your cutting technique as you made more cuts?

4. By the end of the activity, did you feel comfortable using the reciprocating saw?

Notes

Name ______________________ Date ____________ Class __________

Lab Activity 4B

Using a Circular Saw

Objectives

After completing this lab activity, you will be able to:

- Cut lumber using a circular saw.

Text Reference

Chapter 4, *Power Tools*

Materials/Equipment

The following tools and equipment are needed to perform these activities:

- Circular saw
- Tape measure
- Speed square or another tool to mark a perpendicular line
- Marker or other marking instrument
- 1×8 lumber, minimum 3′ long (or similar)
- 2×4 lumber, minimum 3′ long (or similar)

Introduction

A circular saw is used to cut framing, sheathing, and other building materials. Although the circular saw is not used for electrical work as frequently as it is for other trades, electricians must be familiar with how to use it.

Safety Review

Follow all safety procedures specified by your instructor and in any appropriate manufacturer's information. The following are additional safety-related items to keep in mind:

- Wear safety glasses when operating a circular saw.
- When changing blades or adjusting blade depth, make sure the circular saw is not connected to electrical power.
- Before using a circular saw or any power tool, be sure you are familiar with the specific features and safety guidelines for the tools. Review the manufacturer's user manual carefully.

Cutting with a Circular Saw

1. Secure the 1×8 board to a workbench or table using clamps. Push and pull the board to make sure it is secured.
2. Use the tape measure to measure and mark 3″ from one end of the board.
3. Use a speed square or similar instrument to draw a line across the face of the board at the 3″ mark. This is your cut line.

4. Prepare the circular saw.
   A. Make sure the circular saw is not connected to power (corded) or detach the battery (cordless).
   B. If a blade is already installed, verify that it is the correct blade type for the type of material and type of cut. Check that the blade is installed correctly and securely. Inspect the blade to ensure it is clean and that the teeth are not dull or damaged.
   C. If the blade must be installed or changed, follow the manufacturer's directions for doing so.
   D. Adjust the depth of the blade so the bottom point on the blade is between 1/8″ to 1/4″ below the bottom face of the material being cut. Set the saw on the board to be cut to check that the blade depth is set correctly.
   E. Check that the depth adjustment and bevel adjustment screws are tight.
   F. Check that the guard moves smoothly and operates properly.
5. Attach the battery (cordless) or plug in the circular saw (corded).
6. Prepare to make the cut.
   A. Cut so that the base of the circular saw rests on the secured part of the board, not on the portion that will fall after the cut.
   B. Visualize the process of making the cut and ensure that there are no obstructions in the path of the circular saw or blade.
   C. Inspect the board along the cut line for knots or other imperfections. These may cause kickback or binding.

SAFETY NOTE

Keep the following in mind when making a cut with a circular saw:

- Always keep two hands on the circular saw when making a cut.
- Keep a firm grip and be prepared for kickback or blade binding.
- If the blade binds, release the trigger immediately and keep the saw flat against the material. Do not lift the saw from the cut until the blade has come to a complete stop.
- Kickback is a sudden movement of the back end of the saw up and back. If kickback occurs, maintain firm control of the saw and immediately release the trigger.
- Never lift the saw out of the cut while the blade is turning. Wait for the blade to come to a complete stop before lifting the saw from the cut.

7. Make the cut, keeping the following in mind:
   A. Fully depress the trigger so the blade is turning at full speed throughout the cut.
   B. The base of the circular saw will have a small notch to align with the cut line on the board. As you make the cut, keep the cut line positioned in the notch to ensure a straight cut.
   C. Cut straight across the board. If the cut veers to the left or right, the blade may bend.
   D. To begin the cut, align the saw with the cut line and ensure that there is a small gap between the blade and the board. Squeeze the trigger to bring to the blade to full speed, and then move the saw and blade into the board.
   E. Push the circular saw through the cut. Continue the cut until the cut piece falls away. Once the piece falls, release the trigger but keep the saw on the board until the blade comes to a complete stop.
   F. After completing the first cut, unplug the circular saw (corded) or detach the battery (cordless).
   G. Repeat this process to make a second and then a third cut. Be sure to work through all checks on the saw, blade, and board before making a cut.
   H. After completing three cuts of the 1×8 board, repeat the procedure to make three cuts from the 2×4 board. Note that you will need to adjust the blade depth to cut the thicker board.

Name ____

Reflect and Analyze

1. What was your level of experience and comfort with using a circular saw prior to completing this activity?

2. After completing this activity, how would you characterize your level of comfort with using a circular saw?

3. Place the three 3″ pieces of 1×8 you cut next to each other. Are the cuts square? Is there any variation in the cut length of the pieces?

4. Place the three 3″ pieces of 2×4 you cut next to each other. Are the cuts square? Is there any variation in the cut length of the pieces?

5. Did you encounter any problem or issue during this activity? If so, please describe.

Notes

Name ______________________ Date ____________ Class ________

Lab Activity 4C

Using an Oscillating Tool

Objectives

After completing this lab activity, you will be able to:

- Cut openings in drywall using an oscillating tool.

Text Reference

Chapter 4, *Power Tools*

Materials/Equipment

The following tools and equipment are needed to perform these activities:

- Wall frame with attached drywall
- Oscillating tool with drywall blade
- Torpedo level
- Outlet box
- Pencil

Introduction

An oscillating tool is a versatile and easy-to-use tool that can be used to make many types of cuts quickly. Blades are selected based on the material being cut.

Safety Review

Follow all safety procedures specified by your instructor and in any appropriate manufacturer's information. The following are additional safety-related items to keep in mind:

- Wear safety glasses when using any type of power tool.
- When making a cut with an oscillating tool, hold the oscillating tool firmly with both hands.
- Oscillating tool blades can become hot during cutting. Allow the blade time to cool after a cut before handling.

Using an Oscillating Tool

In this activity, you will use an oscillating tool to cut three openings in drywall.

1. Determine the location for the opening in the drywall based on your instructor's direction.
2. Determine the stud locations to ensure that the opening will be in a void between studs.
3. Place the outlet box against the drywall.
4. Place the torpedo level on top of the box, and adjust the box as needed to be level.
5. Hold the box firmly against the drywall so the box does not move, move the torpedo level, and use a pencil to trace the outside of the box.
6. Remove the outlet box from the drywall. Darken the outline if needed.

7. Prepare the oscillating tool.
   A. Detach the battery (cordless) or make sure the tool is unplugged (corded).
   B. Review the manufacturer's user manual to ensure you understand the controls and adjustments.
   C. Inspect the oscillating tool to make sure it is not damaged and appears to be in good working order.
   D. Install a blade appropriate for cutting drywall.
   E. Pull and wiggle the blade to ensure it is securely installed.
   F. Many oscillating tools have trigger locks. If the tool you are using has a trigger lock, familiarize yourself with the locking and unlocking mechanism.
8. Cut the opening with the oscillating tool.
   A. Hold the oscillating tool firmly with both hands while making a cut.
   B. Activate the blade and then slowly bring the blade in contact with the drywall surface. The blade may jump or kick slightly at this initial contact, so be prepared and move the blade as needed to keep it on the cutting line.
   C. Begin the cut slowly and adjust pressure based on how the blade is cutting through the material.
   D. After completing the cut for the first side of the opening, repeat for the other three sides of the opening.
   E. Your instructor may direct you to cut multiple openings in the drywall for additional practice.

Reflect and Analyze

1. Did you encounter any difficulties while completing this activity? If so, please describe.

____

____

____

____

____

2. After completing this activity, what is your level of comfort or confidence using an oscillating tool?

____

____

____

____

____

Name ______________________________ Date ______________ Class __________

Lab Activity 4D

Using a Drill

Objectives

After completing this lab activity, you will be able to:

- Drill holes in wood.

Text Reference

Chapter 4, *Power Tools*

Materials/Equipment

The following tools and equipment are needed to perform these activities:

- Wall frame with exposed studs
- Board or lumber, 1/2″ to 1″ thick, minimum 1 sq. ft. surface
- General-use electric drill
- Right angle drill
- Drill bit, 1/8″ to 1/2″
- Hole saw, 3/4″ to 2″
- Auger bit, 3/4″
- Scratch awl or nail punch
- Tape measure, square, or rule
- Pencil

Introduction

Drilling holes is a common activity for electricians. Holes in framing members and other building materials are needed to run cable or conduit through a structure.

On an electric drill, the part that holds the drill bit is called the chuck. Chucks may be keyed or keyless, and are typically sized at 3/8″ or 1/2″. The size of the chuck is the size of the maximum drill bit shank that will fit in the chuck.

Most drills are reversible, rotating the drill bit clockwise to drill holes and counterclockwise to remove the drill bit from the material after completing a hole.

Safety Review

Follow all safety procedures specified by your instructor and in the appropriate manufacturer's information. The following are additional safety-related items to keep in mind:

- Wear safety glasses when operating any power tool.
- When drilling holes, hold the drill firmly in both hands.
- Do not push the drill with too much pressure. Only light pressure is needed when drilling because the drill bit pulls itself into the hole as it rotates.

Drilling Small Holes

1. Secure a board or piece of lumber to a workbench or table using clamps.
2. Using a tape measure, square, or rule, place a mark at 2″, 4″, and 6″ from one corner. Measuring from the same corner but along the other edge, place a mark at 2″, 4″, and 6″. Draw lines from each mark to create a 2″ grid.
3. At the nine points where the grid lines intersect, use the scratch awl or nail punch to make an indentation in the surface of the wood.

4. Prepare the general-use electric drill.
   A. Detach the battery (cordless) or make sure the drill is not connected to power (corded).
   B. Review the manufacturer's user manual and familiarize yourself with the controls on the drill.
   C. Select a 1/8″ to 1/2″ drill bit appropriate for wood.
   D. Install the drill bit in the chuck. Be sure that the chuck is fully tightened so the drill bit does not slip while drilling.
   E. Set the appropriate direction for drilling.
5. Attach the battery (cordless) or plug in the drill (corded).
6. Position the tip of the drill bit in one of the indentations on the board. Hold the drill firmly in both hands. Position the drill so the drill bit is perpendicular (not angled) to the face of the board.
7. Drill the hole.
   A. Begin with a slow drill speed until the hole is started, and then increase to full speed.
   B. If the motor slows while drilling, reduce your pushing force.
   C. If the drill binds, release the trigger to stop the drill. Reverse the direction of the drill and back the drill out. Then, change the direction for drilling and attempt to finish the hole.
   D. As the drill reaches the end of the hole, decrease the speed of the drill to finish the hole.
   E. After completing the hole, reverse the direction of the drill and run the drill while you pull the drill bit out of the hole.
8. Repeat this process to drill the other eight holes at the indented points.

Using a Hole Saw

1. Mark three points on your board that are at least 2″ from the edges and at least 6″ away from one another.
2. Use the scratch awl or nail punch to create an indentation at each of the three points.
3. Select a hole saw (3/4″ to 2″) and install it in the chuck. Be sure to follow the guidelines in Step 4 of the previous process to change the drill bit and inspect the drill.
4. Position the point of the drill bit in the hole saw in one of the three indented points. Hold the drill firmly in both hands. Position the drill so the drill bit is perpendicular (not angled) to the face of the board.
5. Drill the hole using the guidelines provided in Step 7 of the previous procedure.
6. Repeat this process to drill the other two holes at the indented points.

Drilling Holes in Studs

In this activity, you will use a right-angle drill to drill a hole in a stud.

1. Use a tape measure to mark the center of the stud. By drilling in the center, you will have adequate clearance on both sides of the stud.
2. Use the scratch awl or nail punch to create an indentation at the mark.
3. Prepare the right-angle drill.
   A. Detach the battery (cordless) or make sure the drill is not connected to power (corded).
   B. Review the manufacturer's user manual and familiarize yourself with the controls on the drill.
   C. Select a 3/4″ auger bit appropriate for wood.
   D. Install the auger bit in the chuck. Be sure that the chuck is fully tightened so the drill bit does not slip while drilling.
   E. Set the appropriate direction for drilling.
   F. Attach the battery (cordless) or plug in the drill (corded).

Name ____

4. Position the tip of the auger bit in the indentation in the stud. Hold the drill firmly in both hands. Position the drill so the drill bit is perpendicular (not angled) to the face of the stud.
5. Drill the hole.
   A. Begin with a slow drill speed until the hole is started, and then increase to full speed.
   B. If the motor slows while drilling, reduce your pushing force.
   C. If the drill binds, release the trigger to stop the drill. Reverse the direction of the drill and back the drill out. Then change the direction for drilling and attempt to finish the hole.
   D. As the drill reaches the end of the hole, decrease the speed of the drill to finish the hole.
   E. After completing the hole, reverse the direction of the drill and run the drill while you pull the drill bit out of the hole.

Reflect and Analyze

1. Did you encounter any problems while completing this activity? If so, please describe.

____

____

____

____

____

2. Which was the most difficult type of hole to drill in this activity? Why?

____

____

____

____

____

3. How would you describe your comfort level drilling different types of holes after completing this activity?

____

____

____

____

____

Notes

Name ______________________________ Date ______________ Class ____________

Lab Activity 5A

Using a Digital Multimeter

Objectives

After completing this lab activity, you will be able to:

- Measure voltage, resistance, and continuity with a digital multimeter.

Text Reference

Chapter 5, *Electrical Test Equipment*

Materials/Equipment

The following tools and equipment are needed to perform these activities:

- Digital multimeter (DMM)
- Receptacle, installed
- Single-pole switch
- Piece of insulated conductor
- Light bulb
- Panelboard

Introduction

Electricians use electrical test equipment to measure electrical characteristics in circuits and devices. These measurements can be used to ensure a circuit is safe to work on, to troubleshoot circuit and device problems, and to check that a circuit is functioning correctly.

Safety Review

Follow all safety procedures specified by your instructor and in any appropriate manufacturer's information. The following is an additional safety-related item to keep in mind:

- Be careful when performing electrical tests on live circuits. Be deliberate in your actions and avoid making unnecessary contact with exposed terminals and conductors.

Using a Digital Multimeter

1. Examine the selector dial on your digital multimeter (DMM). List the different types of measurements available and the range settings for each type of measurement.

____

____

____

____

____

____

____

____

Measuring Voltage

1. Set the DMM selector dial to measure ac voltage. Select the highest range.
2. Insert the black lead into the port labeled COM and red lead into the port labeled V.

> **SAFETY NOTE**
> When taking measurements with a digital multimeter, keep your hands on the insulated part of the probes. Never touch the metal portion of a probe while taking a measurement.

3. Using a 120 V receptacle identified by your instructor, take the voltage measurements listed below. If the initial reading is too small to display at the highest range setting, switch to lower range.

   Upper part of duplex receptacle:

   A. Voltage between long opening and short opening: ____________________

   B. Voltage between long opening and U-shaped opening: ____________________

   C. Voltage between short opening and U-shaped opening: ____________________

   Lower part of duplex receptacle:

   A. Voltage between long opening and short opening: ____________________

   B. Voltage between long opening and U-shaped opening: ____________________

   C. Voltage between short opening and U-shaped opening: ____________________

Measuring Resistance

1. Position the DMM selector dial to measure resistance. Select the highest range.
2. Insert the black lead into the port labeled COM and red lead into the port labeled Ω.
3. Obtain a single-pole switch device (with no conductors connected to it) and place the switch in the ON (up) position.
4. Measure the resistance between the two switch terminals. Again, if the initial range selection is too large, select a smaller range setting.

   A. Resistance, switch in ON position: ____________________

5. Place the switch in the OFF (down) position and measure the resistance between the two switch terminals:

   A. Resistance, switch in OFF position: ____________________

6. Obtain a light bulb. Measure the resistance of the light bulb. On a standard light bulb, the portion of the bulb that is screwed into the socket is called the cap. To measure resistance, place one probe at the tip of the cap and the other probe on the side of the cap.

   A. Resistance, light bulb: ____________________

Measuring Continuity

When measuring continuity, the testing device determines if there is a complete circuit path from one probe location to the other. Continuity checks determine if there is an open or break along a conductor or circuit path.

Not all digital multimeters include a continuity setting. In these cases, a resistance measurement can be used to test for continuity. If a reasonably low resistance is present, the path is continuous. If the resistance is infinite or OL, the path is open and continuity is not present.

1. Position the DMM selector dial to measure continuity (or resistance at the highest range if continuity is not an option).

Name ____

2. Insert the black lead into the port labeled COM and red lead into the port labeled Ω.
3. Obtain an insulator conductor. The length of the conductor does not matter. Check for continuity (or resistance) by attaching one probe to one end of the conductor and the other probe to the other end of the conductor.

   A. Is continuity present? ____

Reflect and Analyze

1. Did you encounter any problems or issues while completing this activity? If so, please describe.

____

____

____

____

____

2. Are the voltage measurements taken at the receptacle the values you expected? If not, which measurement was unexpected and why?

____

____

____

____

____

3. Are the resistance measurements taken at the switch and light bulb the values you expected? If not, which measurement was unexpected and why?

____

____

____

____

____

4. Was the continuity test result your expected result? If not, what did you expect and why?

____

____

____

____

____

Notes

Name ______________________ Date ______________ Class ____________

Lab Activity 5B

Using Electrical Test Equipment

Objectives

After completing this lab activity, you will be able to:

- Check for voltage with a noncontact voltage tester.
- Measure current with a clamp meter.
- Check receptacle wiring with a receptacle tester.
- Determine a device's circuit with a circuit breaker finder.

Text Reference

Chapter 5, *Electrical Test Equipment*

Materials/Equipment

The following tools and equipment are needed to perform these activities:

- Noncontact voltage tester
- Clamp meter
- Receptacle tester
- Circuit breaker finder
- Two receptacles, installed
- Single-pole switch, installed
- Panelboard

Introduction

Electricians use electrical test equipment to measure electrical characteristics in circuits and devices. These measurements can be used to ensure a circuit is safe to work on, to troubleshoot circuit and device problems, and to check that a circuit is functioning correctly.

Safety Review

Follow all safety procedures specified by your instructor and in any appropriate manufacturer's information. The following is an additional safety-related item to keep in mind:

- Be careful when performing electrical tests on live circuits. Be deliberate in your actions and avoid making unnecessary contact with exposed terminals and conductors.

Using a Noncontact Voltage Tester

In this activity, you will use a noncontact voltage tester to test the conductors attached to a receptacle and a single-pole switch.

SAFETY NOTE

In this activity, you will be testing a live receptacle and switch. Do *not* touch any terminals or conductors.

1. Begin with an installed receptacle identified by your instructor. If the receptacle is attached to the box, begin by removing the screws in the receptacle yoke. Pull the receptacle an inch or two out of the box so you can access the conductors.

2. The receptacle will have three conductors attached to it: a black conductor, a white conductor, and a bare or green conductor. Activate the noncontact voltage tester and place the tip of the tester against the insulation of each conductor. Indicate whether voltage is present at each conductor:

   A. Black conductor: ______________________________

   B. White conductor: ______________________________

   C. Bare/green conductor: ______________________________

3. If the receptacle was attached to the box, reattach the receptacle.

4. Access an installed single-pole switch controlling a light. If the switch is attached to the box, begin by removing the screws in the switch yoke. Pull the switch an inch or two out of the box so you can access the conductors.

5. The switch will have three conductors attached to it: a bare or green conductor attached to a green grounding terminal, and two conductors attached to the two switching terminals. Place the switch in the OFF (down) position. Activate the noncontact voltage tester and place the tip of the tester against the insulation of each conductor. Indicate whether voltage is present at each conductor:

   A. Upper switching conductor: ______________________________

   B. Lower switching conductor: ______________________________

   C. Bare/green conductor: ______________________________

6. Place the switch in the ON (up) position. Indicate whether voltage is present at each conductor:

   A. Upper switching conductor: ______________________________

   B. Lower switching conductor: ______________________________

   C. Bare/green conductor: ______________________________

7. If the switch was attached to the box, reattach the switch.

Using a Clamp Meter

In this activity, you will use a clamp to measure current in a conductor.

SAFETY NOTE

In this activity, you will be testing a live switch. Do *not* touch any terminals or conductors.

1. Begin with the single-pole light switch used in the previous activity. If the switch is attached to the box, begin by removing the screws in the switch yoke. Pull the switch an inch or two out of the box so you can access the conductors.

2. Place the clamp meter around one of the switching conductors. Place the switch in the ON position and record the reading on the clamp meter. Then place the switch in the OFF position and record the reading on the clamp meter:

   A. Current (ON position): ______________________________

   B. Current (OFF position): ______________________________

Name ______________________________

Using a Receptacle Tester

In this activity, you will use a receptacle tester to test two duplex receptacles.

1. Activate the receptacle tester and test both plugs in two duplex receptacles identified by your instructor. Record the result provided by the tester:

   Receptacle #1:

   A. Upper portion: ______________________________

   B. Lower portion: ______________________________

   Receptacle #2:

   A. Upper portion: ______________________________

   B. Lower portion: ______________________________

Using a Circuit Breaker Finder

1. Review the manufacturer's user manual to become familiar with the options and proper use of the specific circuit breaker finder you are using.
2. Plug the circuit breaker finder transmitter device into a receptable specified by your instructor.
3. Go to the panelboard containing the circuit breaker. Place the tip of the circuit breaker finder detection device near the top-left circuit breaker. Slowly move the detection device down over each circuit breaker. Repeat this process with the circuit breakers on the right side if necessary. The detection device will indicate when it reaches the circuit breaker for the circuit containing the receptacle being traced.

Reflect and Analyze

1. Did you encounter any problems or issues while completing this activity? If so, please describe.

____

____

____

____

2. What wiring issues can be identified by the receptacle tester you used in this activity?

____

____

____

____

3. If you needed to identify the circuit breaker for a receptacle and did not have a circuit breaker finder, how could you determine the correct circuit breaker?

____

____

____

____

Notes

Name ______________________ Date ______________ Class __________

Lab Activity 7A

Ganging Metal Boxes

Objectives

After completing this lab activity, you will be able to:

- Gang metal boxes.

Text Reference

Chapter 7, *Electrical Boxes*

Materials/Equipment

The following tools and equipment are needed to perform these activities:

- Two gangable metal boxes
- Screwdriver

Introduction

Some metal boxes can be joined together to create a larger box. These boxes are called gangable boxes. To join the boxes, you simply remove one side plate from each box and connect the boxes together.

Safety Review

Follow all safety procedures specified by your instructor and in any appropriate manufacturer's information. The following is an additional safety-related item to keep in mind:

- Metal boxes can have sharp edges and corners. Use caution when handling metal boxes and box side plates.

Ganging Metal Boxes

1. Obtain two gangable metal boxes.
2. Remove side plate set screw for the left side plate on one of the boxes.
3. Slide the side plate out of the box. On some metal boxes, you need to twist the side plate to remove it.
4. Remove the side plate set screw for the right side plate on the other box.
5. Put the boxes together with their open sides toward the middle. Be sure the boxes are pressed together with no gaps.
6. Install the set screws removed earlier to secure the boxes together.

Reflect and Analyze

1. Did you encounter any problems or issues while completing this activity? If so, please describe.

____

____

____

____

____

2. In what situations would ganging boxes be useful for an electrician?

Name ______________________________ Date ______________ Class __________

Lab Activity 8A

Connecting Conductors

Objectives

After completing this lab activity, you will be able to:

- Connect conductors using wire nuts.
- Connect conductors using push-in connectors.

Text Reference

Chapter 8, *Electrical Conductors and Connectors*

Materials/Equipment

The following tools and equipment are needed to perform these activities:

- Conductors, 10 AWG to 14 AWG, short lengths with stripped ends (10 minimum)
- Wire nuts, multiple sizes
- Push-in connectors, multiple sizes
- Lineman pliers

Introduction

Wire connections are a critical part of a circuit. Incorrect wire connections can lead to an improperly functioning electrical system. More importantly, improper connections can result in sparking and burning, and may eventually cause a fire.

Connecting Two Conductors with a Wire Nut

1. Pick two conductors from the available conductors. Record the following information for each conductor:

   Conductor #1: AWG _____ Solid or stranded? _________

   Conductor #2: AWG _____ Solid or stranded? _________

2. From the available wire nuts, select the appropriate wire nut for the number and size of conductors being connected.

   Color of wire nut selected: _________

3. Using the lineman pliers, twist the bare ends of the two conductors together. Continue twisting until the insulated portions of the conductors are twisted at least twice. Note: Most electricians do not twist the conductors in two-wire wire nut connections. However, this activity specifies twisting to provide additional practice.
4. Place the wire nut over the bare twisted ends of the conductors. Turn the wire nut clockwise to tighten onto the conductors. Turn until the wire nut is completely tightened.
5. Inspect the connection. If any bare conductor is visible outside of the wire nut, unscrew the wire nut and redo the connection. (You may need to adjust the length of stripped conductor.)
6. Hold a conductor in each hand, then try to pull the conductors apart from one another. If a conductor pulls out of the wire nut, redo the connection.
7. Ask your instructor to inspect the connection.

   Instructor approval: _________

8. Remove the wire nut by twisting counterclockwise and return the conductors and wire nuts.

Connecting Three Conductors with a Wire Nut

1. Pick three conductors from the available conductors. Record the following information for each conductor:

   Conductor #1: AWG _____ Solid or stranded? _________

   Conductor #2: AWG _____ Solid or stranded? _________

   Conductor #3: AWG _____ Solid or stranded? _________

2. From the available wire nuts, select the appropriate wire nut for the number and size of conductors being connected.

   Color of wire nut selected: _________

3. Using the lineman pliers, twist the bare ends of the three conductors together. Continue twisting until the insulated portions of the conductors are twisted at least twice.
4. Place the wire nut over the bare twisted ends of the conductors. Turn the wire nut clockwise to tighten onto the conductors. Turn until the wire nut is completely tightened.
5. Inspect the connection. If any bare conductor is visible outside of the wire nut, unscrew the wire nut and redo the connection. (You may need to adjust the length of stripped conductor.)
6. Pull the conductors apart from one another. If a conductor pulls out of the wire nut, redo the connection.
7. Ask your instructor to inspect the connection.

   Instructor approval: _________

8. Remove the wire nut by twisting counterclockwise and return the conductors and wire nuts.

Connecting Four Conductors with a Wire Nut

1. Pick four conductors from the available conductors. Record the following information for each conductor:

   Conductor #1: AWG _____ Solid or stranded? _________

   Conductor #2: AWG _____ Solid or stranded? _________

   Conductor #3: AWG _____ Solid or stranded? _________

   Conductor #4: AWG _____ Solid or stranded? _________

2. From the available wire nuts, select the appropriate wire nut for the number and size of conductors being connected.

   Color of wire nut selected: _________

3. Use the procedure above to connect and inspect the four-wire connection.
4. Ask your instructor to inspect the connection.

   Instructor approval: _________

5. Remove the wire nut by twisting counterclockwise and return the conductors and wire nuts.

Connecting Conductors with a Push-In Connector

1. Pick four conductors from the available conductors. Record the following information for each conductor:

   Conductor #1: AWG _____ Solid or stranded? _________

   Conductor #2: AWG _____ Solid or stranded? _________

   Conductor #3: AWG _____ Solid or stranded? _________

   Conductor #4: AWG _____ Solid or stranded? _________

Name ____

2. Select an appropriate push-in connector for the number and size of conductors being connected.

   Color of push-in connector selected: ________

3. Follow the push-in connector manufacturer's procedure for making the connection.
4. Ask your instructor to inspect the connection.

   Instructor approval: ________

5. Remove the conductors from the push-in connector and return the conductors and connectors.

Reflect and Analyze

1. Did you encounter any problems or issues while completing this activity? If so, please describe.

   ____
   ____
   ____
   ____
   ____

2. Compare making a connection with a wire nut with making a connection using a push-in connector.

   ____
   ____
   ____
   ____
   ____

3. For a professional electrician, what do you think are some benefits and disadvantages of using wire nuts compared with using push-in connectors?

   ____
   ____
   ____
   ____
   ____

Notes

Name ______________________ Date ______________ Class __________

Lab Activity 9A

Cutting and Bending Electrical Metallic Tubing (EMT)

Objectives

After completing this lab activity, you will be able to:

- Cut electrical metallic tubing (EMT).
- Perform a 90° stub-up bend.
- Perform a back-to-back bend.

Text Reference

Chapter 9, *Electrical Conduit*

Materials/Equipment

The following tools and equipment are needed to perform these activities:

- One 10′ length of 1/2″ EMT
- Tape measure
- Marker or marking instrument
- Hacksaw
- Pipe cutter
- Conduit reamer
- Hand bender (1/2″)

Introduction

Most residential branch circuits are installed using NM cable. However, in some instances, such as when additional protection is needed, conduit may be required. In this activity, you will learn to cut electrical metallic tubing (EMT) and to perform simple conduit bends.

Safety Review

Follow all safety procedures specified by your instructor and in any appropriate manufacturer's information. The following are additional safety-related items to keep in mind:

- Wear safety glasses when working with any tool.
- When cutting conduit, cut edges may have sharp burrs. Handle cut conduit carefully.

Cutting Electrical Metallic Tubing

1. Use the tape measure to mark the tubing 8″ from one end.
2. Secure the tubing in a pipe vise or other holding device.
3. Use the hacksaw to cut the tubing at the mark.
4. Use a conduit reamer to remove any sharp edges and burrs along the cut surfaces.
5. Repeat Steps 1–4 to cut a second 8″ piece with the hacksaw.

6. Measure and cut two more 8″ lengths using a pipe cutter. To use the pipe cutter:
   A. Place the pipe cutter at the mark.
   B. Tighten the pipe cutter until the blade is pressing into the tubing.
   C. Spin the cutter around the tubing.
   D. Tighten the blade and spin the cutter to deepen the cut.
   E. Continue tightening and turning until the tubing is completely cut.
7. Use a conduit reamer to remove any sharp edges and burrs along the cut surfaces.

Bending a 90° Stub-Up

After completing the previous activity, you now have four 8″ lengths of EMT and one piece approximately 7′4″ in length. In this part of the activity, you will make an 18″ stub-up at one end of the long piece of tubing.

1. Find the bend radius of your hand bender. This will be marked or stamped on the bender.
2. Subtract the bend radius from the stub-up distance to determine the distance to mark. For example, if the bend radius is 5″ and the stub-up distance is 18″, the distance to mark is 13″ (18 – 5).
3. Use the tape measure to mark the distance mark from the end of the tubing.
4. Align the arrow on the hand bender with the mark on the tubing. Refer to **Figure 9A-1**. Be sure the bender is positioned to bend the tubing in the proper direction.

Goodheart-Willcox Publisher

Figure 9A-1. Align the arrow on the hand bender with the mark on the tubing.

5. Make the bend using the following guidelines:
   A. Place one foot with your heel on the floor and your toes on the tubing (to hold the tubing in place).
   B. Place your other foot on the bender.
   C. Hold the bender's handle with both hands near the end of the handle.
   D. While pushing down with your foot, pull the bender with your hands.
   E. Bend in a smooth motion.
6. When the bend appears to be at 90°, use a torpedo level to check. Adjust the bend as needed to obtain 90°.

Name ______________________________

Making a Back-to-Back Bend

In the previous portion of the activity, you created an 18″ stub-up at one end of your tubing. Now you will create a back-to-bend and add a 20″ stub-up on the other end of the tubing. The distance between the outside edges of the two stub-ups is 42″.

1. Beginning at the outside edge of the stub-up, measure and mark the distance between stub-ups (42″) on the tubing.
2. Align the star on the hand bender with the mark on the tubing. Refer to **Figure 9A-2**. Be sure the bender is positioned to bend the tubing in the proper direction.

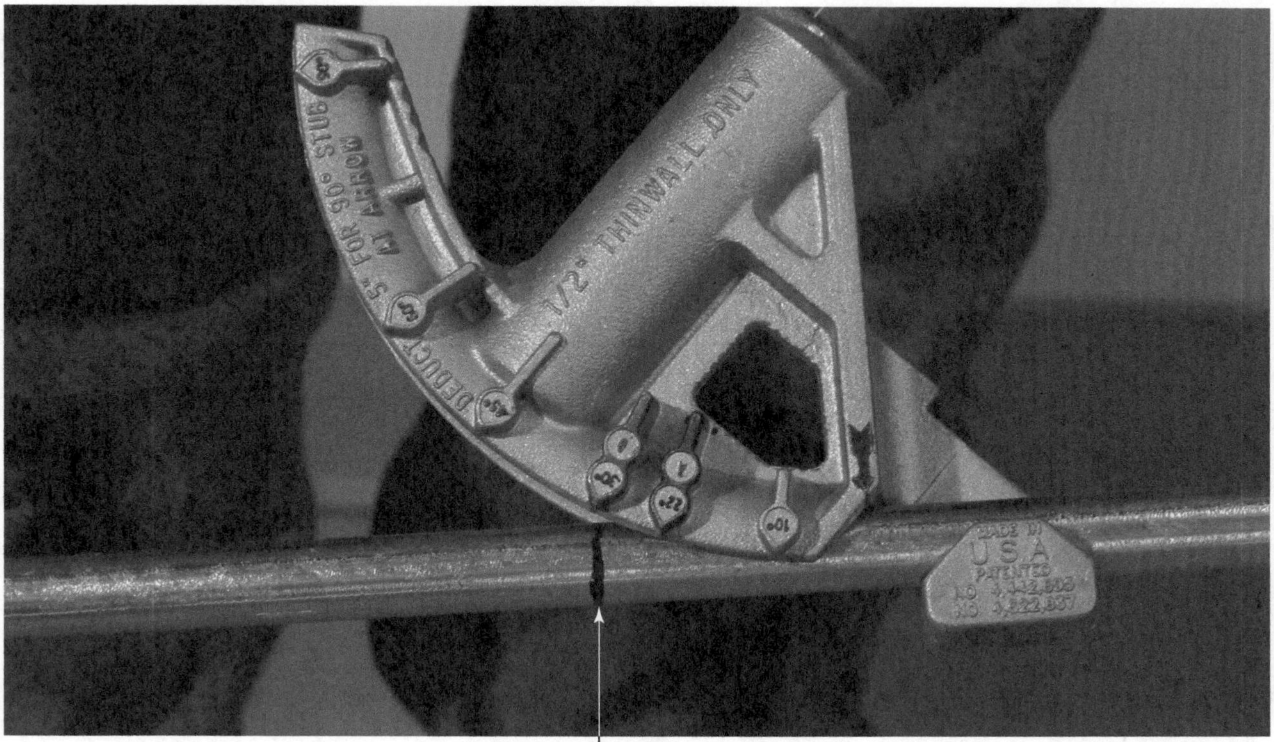

Mark on EMT aligned with star on hand bender

Goodheart-Willcox Publisher

Figure 9A-2. Align the star on the hand bender with the mark on the tubing.

3. Make the bend using the following guidelines:
   A. Place one foot with your heel on the floor and your toes on the tubing (to hold the tubing in place).
   B. Place your other foot on the bender.
   C. Hold the bender's handle with both hands near the end of the handle.
   D. While pushing down with your foot, pull the bender with your hands.
   E. Bend in a smooth motion.
4. When the bend appears to be at 90°, use a torpedo level to check. Adjust the bend as needed to obtain 90°.
5. From the bottom of the horizontal portion of conduit, measure and mark a distance of 20″ on the second stub-up.
6. Cut the second stub-up to height using a hacksaw or pipe cutter.
7. Use a conduit reamer to smooth the cut edges of the tubing.

Reflect and Analyze

1. Did you encounter any problems or issues while completing this activity? If so, please describe.

2. Compare cutting EMT with a hacksaw and cutting EMT with a pipe cutter. Which method was faster? Which produced the cleanest cut? Which do you prefer and why?

Name ______________________ Date ____________ Class __________

Lab Activity 13A

Basic Switched Circuits

Objectives

After completing this lab activity, you will be able to:

- Install a single-pole switch to control a light.
- Install three-way switches to control a light.
- Install three-way and four-way switches to control a light.

Text Reference

Chapter 13, *Basic Switched Circuits*

Materials/Equipment

The following tools and equipment are needed to perform these activities:

- Switch boxes (at least three)
- Single-pole switch
- Three-way switches (at least two)
- Four-way switch
- Light fixture box
- Light fixture
- Length of 14/2 and 14/3 cable (at least two of each)
- Wire connectors
- Pigtails

Introduction

In this activity, you will create various types of switched circuits using three switch boxes and one lighting box.

Safety Review

Follow all safety procedures specified by your instructor and in any appropriate manufacturer's information. The following are additional safety-related items to keep in mind:

- Wear safety glasses when working with tools.
- Before wiring any devices, de-energize the circuit and follow any lockout requirements. Test that voltage is not present before touching any wires.

Preparing to Wire Circuits

This portion of the activity can be skipped if the switch boxes and lighting box are already installed.

1. Install three switch boxes and one lighting box in a wood frame wall or similar as directed by your instructor. The boxes must be close enough that lengths of 14/2 and 14/3 cable can reach from box-to-box comfortably.

Single-Pole Switch with Power at the Switch

1. Make sure that all cables are de-energized.
2. Use the cables, pigtails, wire connectors, switch, and lighting fixture to wire the switched circuit shown in **Figure 13A-1**.
3. Ask your instructor to inspect the installation.

   Instructor approval: _________
4. After the circuit is approved by your instructor, energize the circuit. Does the switching work correctly? If not, what is the problem?

   ____

   ____

   ____

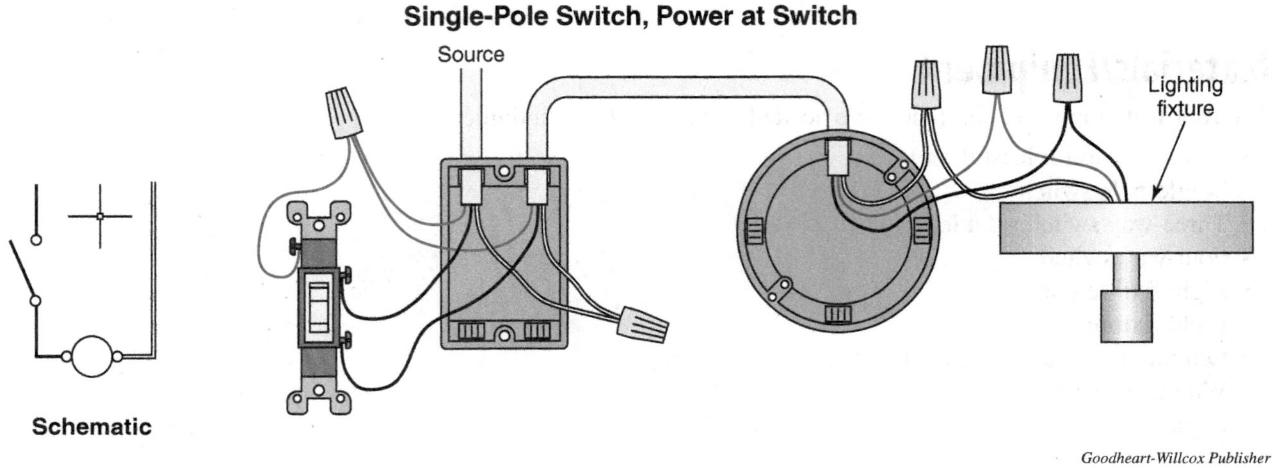

Figure 13A-1. Single-pole switch with power at the switch.

Single-Pole Switch with Power at the Light

1. Make sure that all cables are de-energized.
2. Use the cables, pigtails, wire connectors, switch, and lighting fixture to wire the switched circuit shown in **Figure 13A-2**.
3. Ask your instructor to inspect the installation.

   Instructor approval: _________
4. After the circuit is approved by your instructor, energize the circuit. Does the switching work correctly? If not, what is the problem?

   ____

   ____

   ____

Name ____________________

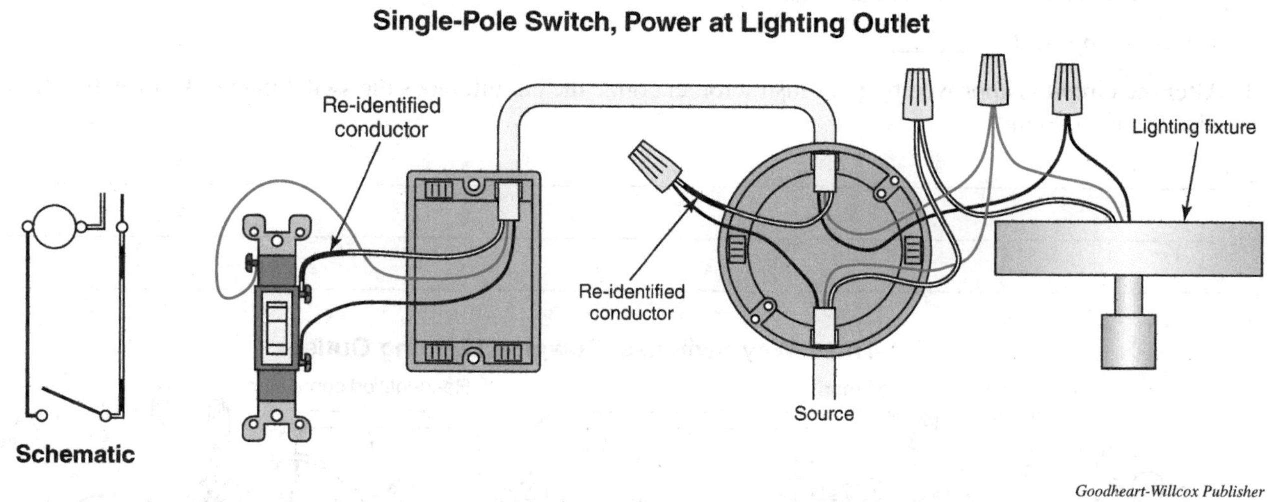

Figure 13A-2. Single-pole switch with power at the light.

Three-Way Switches with Power at a Switch

1. Make sure that all cables are de-energized.
2. Use the cables, pigtails, wire connectors, switches, and lighting fixture to wire the switched circuit shown in **Figure 13A-3**.
3. Ask your instructor to inspect the installation.

   Instructor approval: _________
4. After the circuit is approved by your instructor, energize the circuit. Does the switching work correctly? If not, what is the problem?

   ____

   ____

   ____

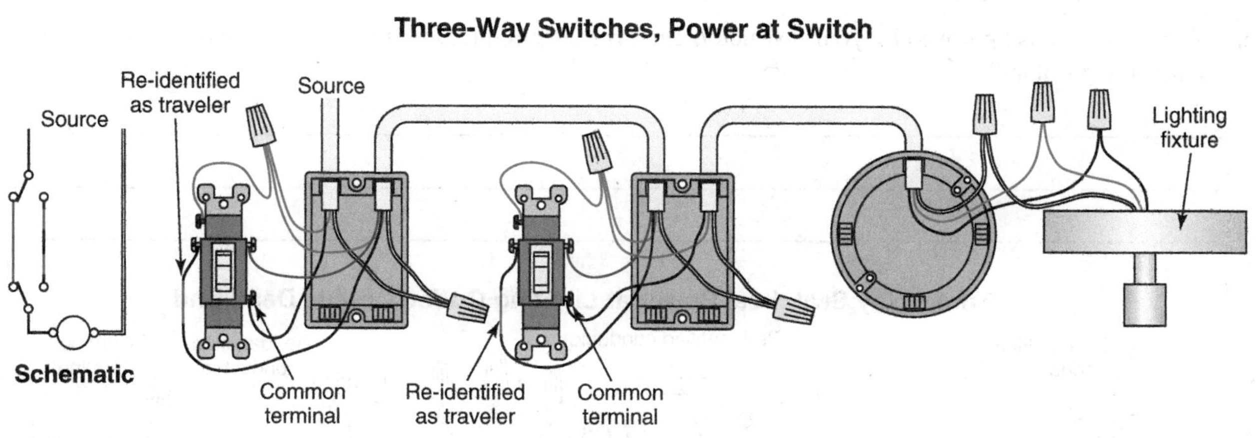

Figure 13A-3. Three-way switches with power at a switch.

Three-Way Switches with Power at the Light

1. Make sure that all cables are de-energized.
2. Use the cables, pigtails, wire connectors, switches, and lighting fixture to wire the switched circuit shown in **Figure 13A-4**.

3. Ask your instructor to inspect the installation.

   Instructor approval: _________

4. After the circuit is approved by your instructor, energize the circuit. Does the switching work correctly? If not, what is the problem?

____

____

____

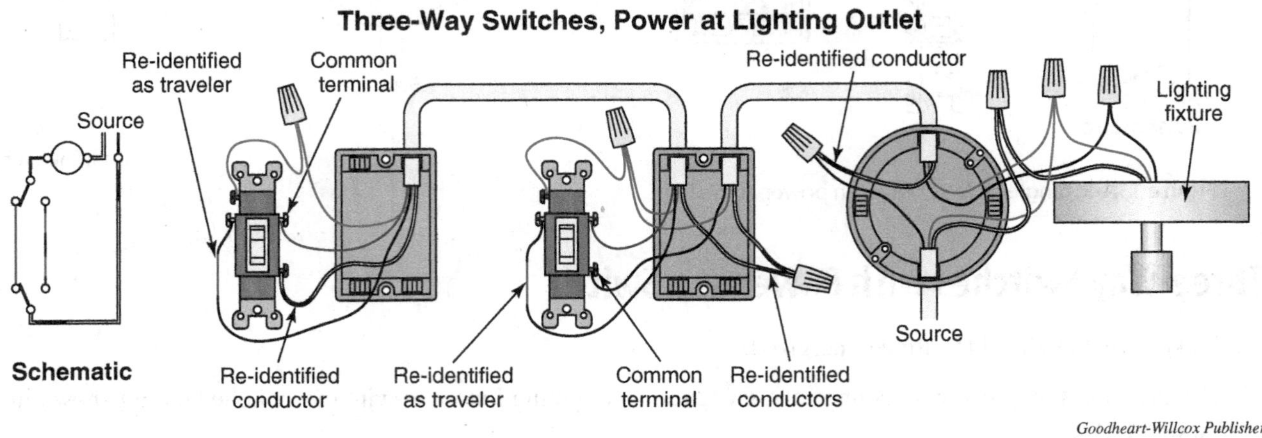

Figure 13A-4. Three-way switches with power at the light.

Three-Way Switches with Power at the Light, Double Dead End

1. Make sure that all cables are de-energized.
2. Use the cables, pigtails, wire connectors, switches, and lighting fixture to wire the switched circuit shown in **Figure 13A-5**.
3. Ask your instructor to inspect the installation.

   Instructor approval: _________

4. After the circuit is approved by your instructor, energize the circuit. Does the switching work correctly? If not, what is the problem?

____

____

____

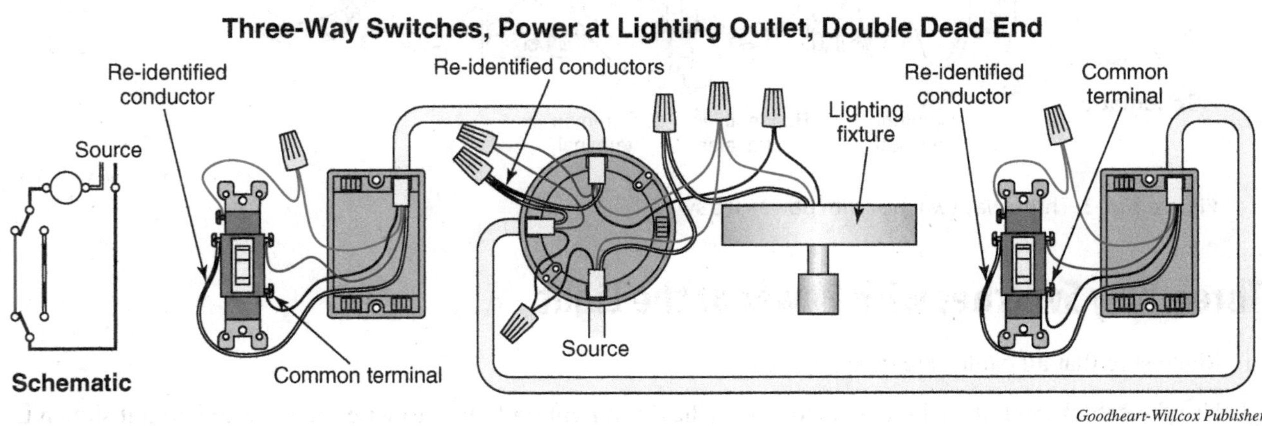

Figure 13A-5. Three-way switches with power at the light, double dead end.

Name ______________________________

Four-Way Switch with Power at the Switch

1. Make sure that all cables are de-energized.
2. Use the cables, pigtails, wire connectors, switches, and lighting fixture to wire the switched circuit shown in **Figure 13A-6**.
3. Ask your instructor to inspect the installation.

   Instructor approval: _________
4. After the circuit is approved by your instructor, energize the circuit. Does the switching work correctly? If not, what is the problem?

   ____

   ____

   ____

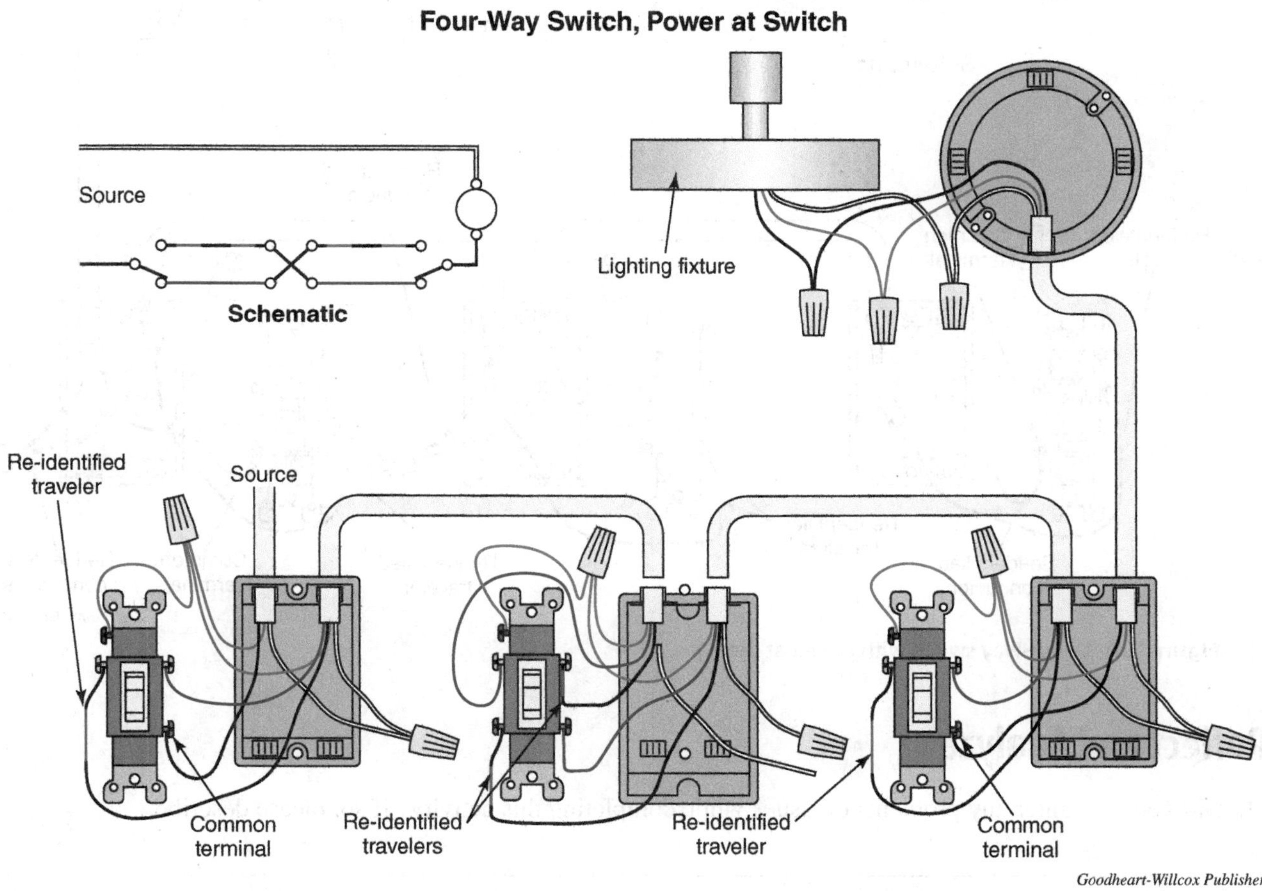

Figure 13A-6. Four-way switch with power at the switch.

Four-Way Switch with Power at the Light

1. Make sure that all cables are de-energized.
2. Use the cables, pigtails, wire connectors, switches, and lighting fixture to wire the switched circuit shown in **Figure 13A-7**.
3. Ask your instructor to inspect the installation.

   Instructor approval: _________

4. After the circuit is approved by your instructor, energize the circuit. Does the switching work correctly? If not, what is the problem?

____

____

____

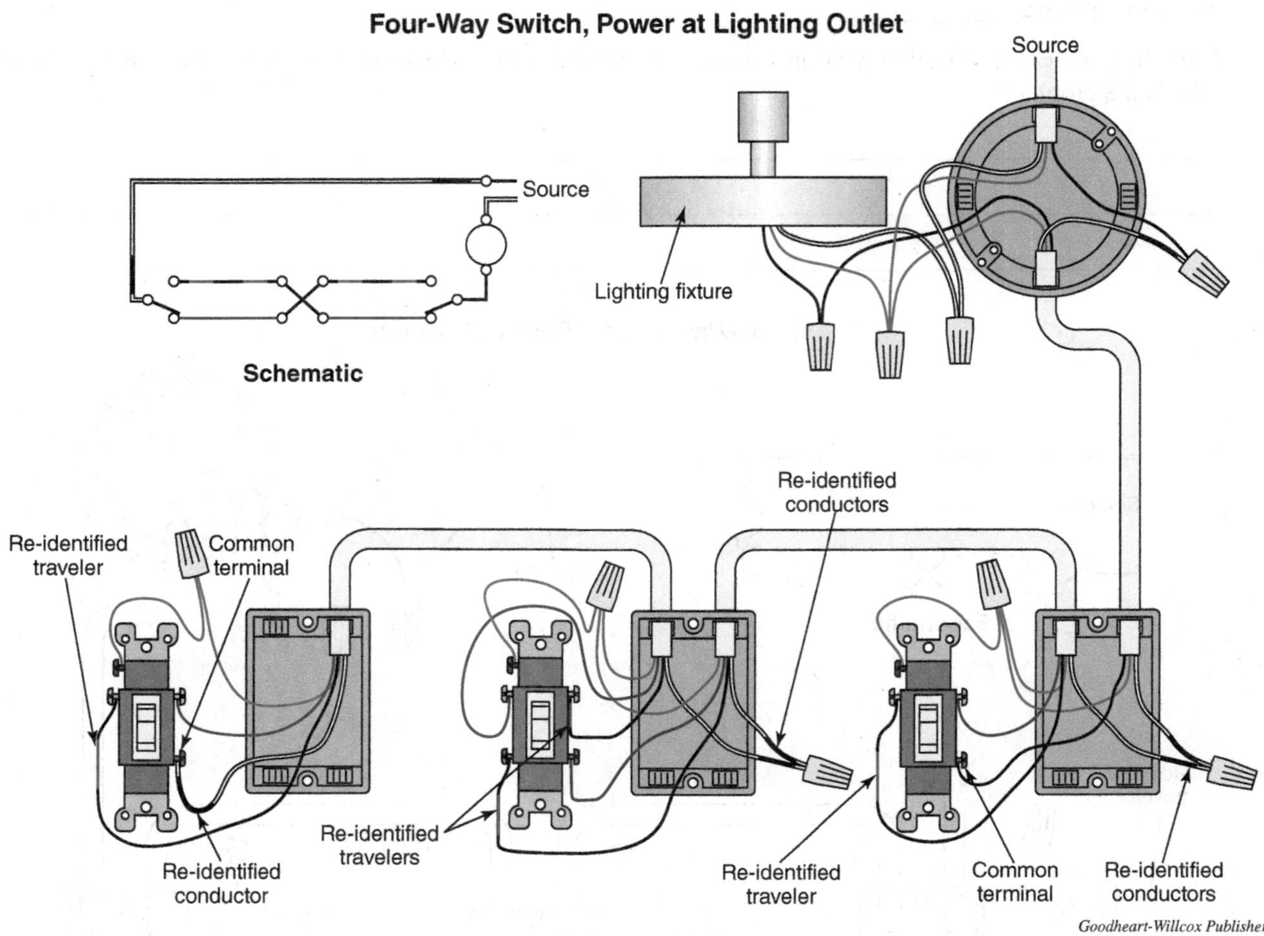

Figure 13A-7. Four-way switch with power at the light.

Reflect and Analyze

1. Did you encounter any problems or issues while completing this activity? If so, please describe.

____

____

____

____

____

Name ______________________________

2. How are four-way switches similar to three-way switches? How are the two types of switches different?

3. If you needed to add a fourth switch to a lighting circuit, how would you add it?

Notes

Name ______________________________ Date ______________ Class __________

Lab Activity 18A

Planning Branch Circuits

Objectives

After completing this lab activity, you will be able to:

- Plan receptacle locations to meet *NEC* requirements.
- Plan lighting locations to meet *NEC* requirements.
- Plan branch circuits to meet *NEC* requirements.

Text Reference

Chapter 18, *Branch Circuit/Receptacle Requirements*

Materials/Equipment

The following tools and equipment are needed to perform these activities:

- Paper and pencil
- Residential floor plan drawing with dimensions

Introduction

In this activity, you will design portions of a residential electrical system following the requirements found in the *NEC*.

Activity

1. Obtain a residential floor plan that shows room sizes or dimensions. This floor plan may be provided by your instructor, or you may need to find a floor plan on the internet.
2. Create a sketch of the floor plan and identify the locations of receptacles, lighting fixtures, and switches:
   - Identify receptacles that require AFCI and/or GFCI protection.
   - Connect switch symbols to the corresponding lighting fixture symbol.
   - Connect outlets that are on the same circuit to one another.
   - Note the current rating of each circuit.
   - Include exterior lighting fixtures and receptacles.
3. Ask your instructor to review your plan and note any items to be added, eliminated, or adjusted.

____

____

____

____

____

Reflect and Analyze

1. Did you encounter any problems or issues while completing this activity? If so, please describe.

2. Based on your plan, how many circuits will the house require?

3. Are there any electrical circuits or loads that are not included in your plan? If so, please list the missing circuits.